Vanishing Matter and the Laws of Motion

Routledge Studies in Seventeenth-Century Philosophy

1. The Soft Underbelly of Reason
The Passions in the
Seventeenth Century
Edited by Stephen Gaukroger

2. Descartes and Method
A Search for a Method in Meditations
Daniel E. Flage and
Clarence A. Bonnen

3. Descartes' Natural Philosophy
Edited by Stephen Gaukroger,
John Schuster and John Sutton

4. Hobbes and History
Edited by G. A. J. Rogers and
Tom Sorell

5. The Philosophy of Robert Boyle
Peter R. Anstey

6. Descartes
Belief, Scepticism and Virtue
Richard Davies

7. The Philosophy of John Locke
New Perspectives
Edited by Peter R. Anstey

8. Receptions of Descartes
Cartesianism & Anti-Cartesianism in
Early Modern Europe
Edited by Tad M. Schmaltz

9. Material Falsity and Error in Descartes' Meditations
Cecilia Wee

10. Leibniz's Final System
Monads, Matter, and Animals
Glenn A. Hartz

11. Pierre Bayle's Cartesian Metaphysics
Rediscovering Early Modern
Philosophy
Todd Ryan

12. Insiders and Outsiders in Seventeenth-Century Philosophy
Edited by G.A.J. Rogers, Tom Sorell
and Jill Kraye

13. Vanishing Matter and the Laws of Motion
Descartes and Beyond
Edited by Dana Jalobeanu and
Peter R. Anstey

Vanishing Matter and the Laws of Motion

Descartes and Beyond

Edited by Dana Jalobeanu
and Peter R. Anstey

First published 2011
by Routledge
270 Madison Avenue, New York, NY 10016

Simultaneously published in the UK
by Routledge
2 Park Square, Milton Park, Abingdon, Oxon OX14 4RN

Routledge is an imprint of the Taylor & Francis Group, an informa business

© 2011 Taylor & Francis

The right of the Dana Jalobeanu and Peter R. Anstey to be identified as the authors of the editorial material, and of the authors for their individual chapters, has been asserted in accordance with sections 77 and 78 of the Copyright, Designs and Patents Act 1988.

Typeset in Sabon by IBT Global.
Printed and bound in the United States of America on acid-free paper by IBT Global.

All rights reserved. No part of this book may be reprinted or reproduced or utilised in any form or by any electronic, mechanical, or other means, now known or hereafter invented, including photocopying and recording, or in any information storage or retrieval system, without permission in writing from the publishers.

Trademark Notice: Product or corporate names may be trademarks or registered trademarks, and are used only for identification and explanation without intent to infringe.

Library of Congress Cataloging-in-Publication Data

Vanishing matter and the laws of motion : Descartes and beyond / edited by Dana Jalobeanu and Peter R. Anstey.
    p. cm. — (Routledge studies in seventeenth-century philosophy ; 13)
    Includes bibliographical references and index.
    1. Physics—Philosophy—History—17th century. 2. Physical laws—History—17th century.   I. Jalobeanu, Dana, 1970–   II. Anstey, Peter R., 1962–
    QC7.V36 2011
    530.01—dc22
                                    2010028736

ISBN13: 978-0-415-88266-8 (hbk)
ISBN13: 978-0-203-83338-4 (ebk)

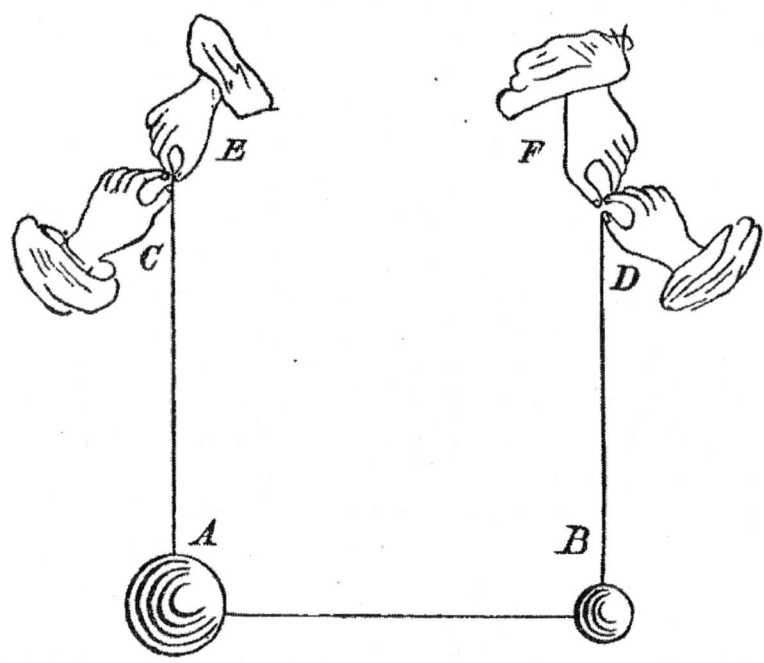

Frontispiece Diagram of Christiaan Huygens's thought experiment for the collision of unequal-sized bodies, Huygens 1888–1950, 6: 341.

Contents

List of Figures ix
List of Abbreviations xi
Acknowledgments xiii

   Introduction 1

PART I
Cartesian Matter

1 The Vanishing Nature of Body in Descartes's Natural Philosophy 11
   MIHNEA DOBRE

2 The New Matter Theory and Its Epistemology: Descartes (and Late Scholastics) on Hypotheses and Moral Certainty 31
   ROGER ARIEW

PART II
Matter, Mechanism, and Medicine

3 Post-Cartesian Atomism: The Case of François Bernier 49
   VLAD ALEXANDRESCU

4 The Matter of Medicine: New Medical Matter Theories in Mid-Seventeenth-Century England 61
   PETER R. ANSTEY

5 Without God: Gravity as a Relational Quality of Matter in Newton's *Treatise* 80
   ERIC SCHLIESSER

PART III
Matter and the Laws of Motion

6 The Cartesians of the Royal Society: The Debate Over
   Collisions and the Nature of Body (1668–1670)   103
   DANA JALOBEANU

7 On Composite Systems: Descartes, Newton, and the
   Law-Constitutive Approach   130
   KATHERINE BRADING

8 Huygens, Wren, Wallis, and Newton on Rules of Impact
   and Reflection   153
   GEMMA MURRAY, WILLIAM HARPER, AND CURTIS WILSON

PART IV
Leibniz and Hume

9 Leibniz, Body, and Monads   195
   DANIEL GARBER

10 Leibniz on Void and Matter   215
    SORIN COSTREIE

11 Hume on the Distinction between Primary and
    Secondary Qualities   235
    JANI HAKKARAINEN

List of Contributors   261
Index   263

Figures

Frontispiece	Diagram of Christiaan Huygens's thought experiment for the collision of unequal-sized bodies, Huygens 1888–1950, 6: 341.	v
1.1	Diagram for *Principles* II, art. 30, AT IXb (rear).	16
1.2	The dissolution of Cartesian bodies: The case of solid bodies.	17
1.3	The dissolution of Cartesian bodies: The case of bodies carried around by other bodies.	20
1.4	*Les Météores*: AT VI 242.	23
8.1	Huygens's diagram examples.	157
8.2	Huygens's basic diagram.	158
8.3	The basic construction.	159
8.4	An example.	159
8a	Huygens 1888–1950, 6: 337.	162
8b	Huygens 1888–1950, 6: 337.	163
8c	Huygens 1888–1950, 6: 338.	164
8d	Huygens 1888–1950, 6: 339.	165
8e	Huygens 1888–1950, 6: 340.	166
8f	Huygens 1888–1950, 6: 341.	167
8g	Huygens 1888–1950, 6: 342.	169
8.5	Wren 1668: 867.	171

8.6	Wren's basic diagram.	173
8.7	Re, Se initial velocities; oR's final velocities.	173
8.8	Ro, So as initial velocities; eR's as final velocities.	174
8.9	Wren's diagram examples exploiting symmetries.	174
8.10	Body 1 hits stationary body 2.	181
8.11	Body 1 overtaking body 2.	181
8.12	A head-on collision.	181
8.13	Newton's diagram with notes added.	183
8.14	Diagram illustrating Newton's calculation.	184
8.15	Pendulum arc velocity diagram.	189
8.16	Euclid applied.	190

Abbreviations

DESCARTES

AT *Œuvres de Descartes*, revised edn, 11 vols, eds. C. Adam and P. Tannery, Paris: Vrin, 1996.

CSM *The Philosophical Writings of Descartes*, trans. J. Cottingham, R. Stoothoff, and D. Murdoch, 2 vols, Cambridge: Cambridge University Press, 1985.

CSMK *The Philosophical Writings of Descartes*, vol. 3, trans. J. Cottingham, R. Stoothoff, and D. Murdoch and A. Kenny, Cambridge: Cambridge University Press, 1991.

HUME

Treatise *A Treatise of Human Nature*, eds. D. F. Norton and M. J. Norton, Oxford: Clarendon Press, 2007; 1st edn 1739–1740.

EHU *An Enquiry concerning Human Understanding*, ed. T. L. Beauchamp, Oxford: Clarendon Press, 2000; 1st edn 1748.

LEIBNIZ

A *Sämtliche Schriften und Briefe*, eds. Deutsche Akademie der Wissenschaften zu Berlin, Berlin: Akademie-Verlag, 1923–. References include series, volume, and page. So "A6.4.1394" is series 6, volume 4, p. 1394. Note that "A2.1^2.123" refers to p. 123 of the second edition of series 2, volume 1.

AG *Philosophical Essays*, eds. and trans. R. Ariew and D. Garber, Indianapolis: Hackett, 1989.

ALC *The Labyrinth of the Continuum: Writings on the Continuum Problem, 1672–1686*, Trans., ed. and comm. R. Arthur, New Haven: Yale University Press, 2001.

G *Die philosophischen Schriften*, 7 vols, ed. C. I. Gerhardt, Berlin: Weidmann, 1875–1890. References include volume and page. So "G VII 80" is volume 7, p. 80.

GM *Leibnizens mathematische Schriften*, 7 vols, ed. C. I. Gerhardt, Berlin: A. Asher, 1849–1863. References include volume and page. So "GM VII 80" is volume 7, p. 80.

L *G. W. Leibniz: Philosophical Papers and Letters*, 2nd edn, ed. and trans. L. E. Loemker, Dordrecht: D. Reidel, 1969.

LC *The Leibniz–Clarke Correspondence*, ed. H. G. Alexander; trans. H. T. Mason, Manchester and New York: Manchester University Press and Barnes & Noble, 1956.

LDB *The Leibniz–Des Bosses Correspondence*, eds. and trans. B. Look and D. Rutherford, New Haven: Yale University Press, 2007.

T *Theodicy. Essays on the Goodness of God, the Freedom of Man, and the Origin of Evil*, ed. A. Farrar; trans. E. M. Huggard, La Salle, Il: Open Court, 1952. Referred to by section, e.g. T 35.

WF *Leibniz's "New system" and Associated Contemporary Texts*, eds. and trans. R. S. Woolhouse, and Richard Francks, Oxford: Oxford University Press, 1997.

LOCKE

Essay *An Essay concerning Human Understanding*, 4th edn, ed. P. H. Nidditch, Oxford: Clarendon Press, 1975; 1st edn 1690.

Acknowledgments

Nine of the eleven chapters included in this volume were presented at a colloquium on "Vanishing bodies and the birth of modern physics" held at New Europe College, Bucharest, in late June and early July 2008. The editors would like to acknowledge the sponsorship of New Europe College, Princeton University, the University of Otago, the University of Notre Dame, and the University of Bucharest. Eric Schliesser's chapter was intended for presentation at the colloquium, but he had to withdraw at the last minute. Dana Jalobeanu's chapter was first presented at the HOPOS conference in Paris, 2006, and appears here for the first time. The editors are also grateful to Sorana Corneanu for her translation from French of Chapter 3.

Illustrations in Chapter 1 from *Œuvres de Descartes*, revised edn, eds. Charles Adam and Paul Tannery, 11 vols, Paris: Vrin, 1996, vol. VI, p. 242 and vol. IXb, p. 79, are reproduced with the permission of Vrin, Paris: copyright Librairie Philosophique J. Vrin, Paris, 1996.

Introduction

Questions about the nature of matter, its qualities, and its behavior were at the very center of early modern philosophy. Robert Boyle spoke for the whole period when he said in his *Excellency of Theology* that

> the Notion of Body in general, *or what it is that makes a thing to be a Corporeal Substance, and discriminates it from all other things*, has been very hotly disputed of, even among the modern Philosophers, & *adhuc sub judice lis est* [and the jury is still out].
> (Boyle 1999–2000, 8: 66)

Indeed, it is not possible to read widely in any major philosopher of the early modern period without encountering these issues. If we restrict ourselves to the traditional philosophical canon, one finds that Descartes's *Principles*, Hobbes's *Leviathan*, Boyle's *Forms and Qualities*, Spinoza's *Ethics*, Locke's *Essay*, Berkeley's *Principles*, and Hume's *Enquiry* all deal with the philosophy of matter and its qualities. Yet the centrality of the philosophy of matter to the early moderns does not exhaust its interest and importance for present-day historians and philosophers of science. For, while many aspects of material substance confounded seventeenth- and eighteenth-century philosophers and natural philosophers alike, many new and exciting explanations of its qualities and behavior were discovered in that period, discoveries that are still taught today in high schools and universities the world over. Every young student of physics today learns Galileo's kinematic laws of motion and Newton's dynamic laws. Every philosophy undergraduate encounters Locke's primary and secondary quality distinction. One cannot progress very far in chemistry without understanding Boyle's Law.

This volume brings together a range of new studies on the philosophy of matter, its qualities and law-like behavior in the early modern period. It does not presume to provide an exhaustive treatment of the subject or even to provide thorough coverage of the issues associated with early modern matter theories. But the studies in this volume are unified by a number of interlocking themes that together enable some of the broader contours of the philosophy of matter to be charted in, what we believe are, new ways.

In the first instance, there is the theme of vanishing matter. Whether we examine the problem of individuation of bodies in Descartes, the emergence of mechanics through the articulation of the laws of motion, or the skeptical trajectory of reflections upon our lack of epistemic access to the inner nature of material objects and of accepting that our ideas are the immediate objects of perception, one finds a tendency for material bodies to vanish. Descartes's bodies vanish in the plenum, Locke denied that we have access to the real essences of things, and Pierre Bayle queried of extension whether "it can very well be reduced to appearances just as colors."[1] The combined impact of such considerations opened up a slippery slope along which the likes of Berkeley and Hume happily slid, and by the time of Boscovich, bodies could be treated as mere points. Several chapters in this volume explore the theme of vanishing bodies, and together they provide a fascinating insight into how the early moderns struggled, not merely to explain matter, but to hold onto it.

A second theme has to do with the laws of nature. It is often supposed that the laws of motion that were discovered in the seventeenth century were predicated upon determinate theories about the quantifiable properties of material bodies. However, some of the studies in this volume reveal that the relation between matter theory and the laws of nature was not so straightforward. From one perspective, it is clear that the varied nature of material bodies provided important empirical evidence for the confirmation of certain laws of motion. But from another perspective, it can be argued that the laws themselves set parameters for determining which qualities could rightly be attributed to material objects in an empirically sound matter theory. The picture that emerges in this volume is one in which the interplay between matter theory and mechanics was far more subtle than is commonly appreciated.

A third and related theme has to do with our knowledge of material bodies and, in particular, the justification of claims about imperceptibly small or distant bodies. What sorts of explanatory principles should constrain our hypotheses about the submicroscopic and what is the epistemic status of these hypotheses? On these points the early moderns had much to say and manifested a broad spectrum of views, views ranging from Descartes's mature moral certainty for inferences to the unobserved to Hume's open and honest skepticism about our prospects of knowledge of the qualities of bodies and even of the bodies themselves.

A fourth theme that emerges from the chapters that follow is that of the diversity of those with an interest in the nature of matter, including those with a vested interest in the direction that the philosophy of matter should take. The remarkable range of interest in matter in the seventeenth century is illustrated as we move from the abstract philosophical speculations of Descartes through the challenges that the new matter theories posed for English physicians mid-century to the careful experiments of the early Royal Society and later Newton, then to the skeptical "armchair reflections" of David Hume. And this is only part of the picture. For, as historians of philosophy know, the theologians were equally interested in the nature of matter and, in particular, in the theory of qualities in so far as it impinged on a number of important Christian doctrines. Little wonder then that it was Descartes's matter theory that was to offend the Parisian theologians and to land his works on the Index in 1663.

Each of these four themes is explored in various ways in the chapters that constitute this volume. Part I is concerned with Cartesian matter theory. In his "The Vanishing Nature of Body in Descartes's Natural Philosophy," Mihnea Dobre argues that in the final analysis, Descartes cannot account for the individuation of bodies in a medium. The Cartesian doctrines of the essence of matter as extension, the infinite divisibility of matter, and his definition of motion render it difficult to specify the necessary and sufficient conditions for the individuation of any body within the plenum. Dobre shows how two later Cartesians, Gerauld de Cordemoy and Jacques du Roure, effectively denied the first two Cartesian doctrines and resorted to a decidedly un-Cartesian commitment to atoms. Dobre also shows how the problem of individuation is only reinforced by a careful analysis of Descartes's own diagrams, which were intended to elucidate his theory.

In fact, the problems for Descartes's matter theory emerge even in his own attempts visually to represent invisible material particles. As early as 1638, he was justifying his inferences from visible to invisible bodies on the basis of a Familiarity Condition, and a similar justification is found in the *Principles*. Interestingly, however, Descartes's views on this kind of transductive inference underwent significant changes as his thought matured. Roger Ariew, in his chapter on the epistemology of the new matter theory, focuses on Descartes's changing views with respect to hypotheses in natural philosophy and, in particular, on the notion of moral certainty. Ariew shows just how widely held the notion of moral certainty was amongst the late Scholastics, thus providing important context for understanding Descartes's own conception. He then uses his analysis of moral certainty in Descartes to show how Descartes's views on the question of the certainty of hypotheses changed from the early confidence in the self-evident foundations of his philosophical project in the *Discourse* of 1637 to his more subtle, even fallibilist, position in the *Principles* of 1644. It has long been appreciated that Descartes's views underwent an important shift with regard to certainty in this period, and Ariew's study succeeds in shedding new light on the these changes. He also indicates the trajectory that reflection on the nature and role of hypotheses in natural philosophy was to take amongst Cartesians and Descartes sympathisers such as Christiaan Huygens.

Cordemoy's and du Roure's abandoning of the Cartesian identification of space with extension did not occur in a vacuum. For Descartes's contemporary Pierre Gassendi had already revived a form of Greek atomism, baptizing it in such a way as to make it compatible with Christianity and using it as a foundation for his version of the mechanical philosophy. Part II of this volume brings together three studies of the new approaches to the theory of matter and its qualities that emerged in the wake of the Cartesian theory and its main atomist competitor. In his study of the post-Cartesian atomism of François Bernier, Vlad Alexandrescu shows how Bernier's reconfiguring and augmentation of Gassendi's defense of atomism included a frontal assault on the Cartesian assumption of the continuity of matter. He also explores the manner in which the traditional Aristotelian primary qualities were explicated by Bernier along atomist lines in a manner entirely consonant with the mechanical philosophy.

Now, it is well known that the kind of Gassendist alternative to Cartesianism that we find in Bernier had made an early impact in Britain in the writings of Walter Charleton (Charleton 1654). However, what is less appreciated is the extent to which this influenced the medical fraternity. As mentioned above, it was not only philosophers and natural philosophers who were interested in the nature of matter in the mid-seventeenth century. New developments in anatomy and iatrochemistry had led to an undermining of the hylomorphic matter theory and its concomitant theory of qualities that underlay the Galenic approach to therapeutic medicine. If the qualities and behavior of matter were not to be explained in terms of the four element theory and *primae qualitates* of Aristotle, then Galen's humoral theory of individual well-being was in need of serious revision or abandonment. It is not surprising, then, that in the 1660s, a debate raged about the efficacy and applicability of the new theory of the qualities of matter that was the primary *explanans* of the corpuscular matter theory of Descartes, Gassendi, and Boyle. In his chapter, Peter Anstey examines the new medical matter theories of the period and provides a survey of the main contours of this debate in the writings of some physicians who may be unfamiliar to historians of philosophy, physicians such as Marchamont Nedham, George Castle, Robert Sprackling, and John Twysden.

The point at issue amongst physicians and natural philosophers alike was the deployment of the mechanical affections of shape, size, motion, and texture, as the new *explanans* in both medicine and natural philosophy. This remained the dominant issue in the development of the theory of the qualities of matter, and by the end of the century it had bedded in as the predominant theory. Controversies over the qualities of matter, however, did not peter out with the broad acceptance of the corpuscular theory of qualities, for the publication of Newton's *Principia* brought a new quality of matter into play, namely, gravity. Eric Schliesser's chapter offers a new interpretation of Newton's early thoughts on the nature of the property of gravity. Schliesser, focusing on the unduly neglected *Treatise of the System of the World* (datable to the 1680s), argues that, for Newton, a lonely corpuscle would not have the property of gravity. This is because the Newton of the *Treatise* viewed gravity as a relational property. Among the happy consequences of attributing this view to Newton is, according to Schliesser, the fact that it is consistent with his subordinating of matter theory to the laws of nature and the manner in which they figure in the discovery of the forces of nature.

Now a determination of the precise relation between Newton's laws and his account of matter is a difficult interpretive issue, and the three chapters that constitute Part III combine to illuminate the variegated nature of this issue. Dana Jalobeanu's analysis of the Royal Society's early debates on the laws of motion (1668–1670) shows how interrelated are the issues of correcting Descartes's mistaken laws of collision, giving a new and more precise account of what counts as a physical body when it comes to formulating the laws of motion, and the epistemology associated with the laws of nature. Exploring a fascinating debate that took place just one decade before the emergence of Newton's theory of motion, Jalobeanu shows that the Royal Society's mathematicians took issue with Descartes's problem of vanishing bodies and struggled to formulate both a more precise philosophy of matter and another set of laws describing the behavior of "mathematically defined" physical bodies. During the debate, the Royal Society was clearly divided between those arguing that an abstract/mathematical definition of physical bodies and an associated "correct" set of laws is enough for describing the very basic phenomena of the mechanical universe and those claiming that such a description is either insufficient or meaningless from the perspective of a sound natural philosophy. In many ways, Jalobeanu's chapter argues, such a debate can be seen as setting the stage for Newton's own problem: how to give an account of bodies in motion without a definition of body and without an associated matter theory.

In her chapter, Katherine Brading develops a complementary approach. Starting from the premise that Newton's approach was a success in terms of giving a full account of bodies as the objects of a mathematical physics, she asks whether a "law-constitutive approach" can give us a philosophy of matter. It would be a different philosophy of matter: an abstract matter theory of a rather different kind than the early seventeenth-century philosophers had in mind. Nevertheless, it would be a theory covering most of the important physical and metaphysical questions concerning the nature of material bodies, the relation between parts and the whole, and the relation between individual objects and the laws describing their behavior.

In their chapter on "Huygens, Wren, Wallis, and Newton on Rules of Impact and Reflection," Gemma Murray, Curtis Wilson, and William Harper investigate the same problematic background of Newton's theory of motion from an even more challenging perspective. They claim both that in developing a new account of the laws of motion Newton is drawing on earlier successes of Huygens, Wallis, and Wren and that his own account has empirical content. The paper provides the first full English translation of a couple of widely cited and often misread natural philosophical papers of the late 1660s and discusses the key points of the conflicting accounts on collisions offered by Huygens, Wallis, and Wren. Then, Newton's celebrated pendulum experiments are introduced and evaluated as providing experimental instances for testing one theory over another while also providing theory-mediated experiments for extending the limited cases covered by his predecessors into a general account of collisions based on the laws of motion.

Part IV of the volume moves back from questions of philosophy of matter *qua* mechanics to the philosophy of matter as foundational for natural philosophical problems in the works of Leibniz and Hume. In his chapter, Daniel Garber summarizes the wide and sometimes bewildering range of Leibnizian philosophies of matter, all seen as tentative answers to the problem of vanishing bodies. Are physical bodies, that is, composite substances, such as the rainbow, aggregates of monads, or mere phenomena, or even a sort of "common dream" of an infinity of monads? Can we make sense of the unity of composite substance by means of a "dominant monad" or a mysterious "substantial bond"? Is there any way to make sense of this baffling diversity? Daniel Garber's proposal is to read Leibniz's attempts to make sense of composite substance as driven by a set of questions very much like those of contemporary particle physics: Given that the ultimate units of reality are something like monads, how can we make sense of the phenomenal world around us?

Some of Leibniz's proposed answers, however, move beyond the realm of physics into mathematics and the celebrated problem of the continuum. For example, if the world is a plenum and bodies are indefinitely divisible composite substances—or worlds within worlds within worlds—how can one determine the boundaries between physics, mathematics, and metaphysics when trying to make sense of the world? In his chapter, Sorin Costreie explores Leibniz's struggle to make sense of these puzzling questions and the way in which his answer shaped Leibniz's position in the celebrated debate with Samuel Clarke and Isaac Newton.

Finally, Jani Hakkarainen's chapter on Hume's account of the primary and secondary quality distinction brings us back to the theory of the qualities of matter. By the mid-eighteenth century, the mechanical affections, Locke's primary qualities, were fully entrenched (along with gravity) as the new *explanans* of the behavior of material bodies. Yet, the tendency of bodies to vanish was once again threatening to undermine the objective status not simply of the secondary qualities of color, taste, and sound, but even the primary qualities themselves. Hakkarainen argues that a careful analysis of Hume's terminology in his discussions of the primary and secondary quality distinction reveals that the old Aristotelian notion of *proper sensibles* was playing a hitherto unrecognised role in Hume's deliberations about the status of the primary qualities. In fact, the logic of Hume's assimilation of the primary qualities to the sensible qualities forces Hakkarainen to the view that Hume is skeptical about the reality of matter itself.

NOTES

1. "Elle peut fort bien être rétuite à l'apparance tout comme les couleurs," Bayle 1740, 4: 543.

BIBLIOGRAPHY

Boyle, R. (1999–2000) *The Works of Robert Boyle*, 14 vols, eds. M. Hunter and E. B. Davis., London: Pickering and Chatto.
Bayle, P. (1740) *Dictionnaire historique et critique*, 4 vols, 5th edn, Amsterdam.
Charleton, W. (1654) *Physiologia Epicuro-Gassendo-Charletoniana*, London.

Part I
Cartesian Matter

1 The Vanishing Nature of Body in Descartes's Natural Philosophy

Mihnea Dobre

Descartes's matter theory, in the form it is presented in the second part of the *Principles of Philosophy*, has been criticized a number of times. His well-known account of material substance in terms of extension and extension alone proved to be very successful in the mechanical explanations employed and yet very problematic in its relation to the fundamental concepts of his physics. This is the case with his arguments in support of his definition of motion, with the consequences concerning a body at rest, and with his explanation of the individuation of bodies. Equally contested by his contemporaries (Cordemoy, Leibniz) and recent scholars (Garber, Grosholz), Descartes's theory of matter is nevertheless very important for the foundation of his physics. As Stephen Gaukroger observes, "Part II of the *Principles* deals with the foundational principles of Descartes' physical theory, which take the form of a synthesis of matter theory and mechanics" (Gaukroger 2003: 93).[1] By *Descartes's theory of matter*, I am referring here to his account that starts with the proof of the existence of bodies and goes through his discussion of the essence of matter up to the "proper" definition of motion.

In a letter from February 5, 1649, Descartes wrote to Henry More, "I do not agree with what you very generously concede, namely that the rest of my opinions could stand even if what I have written about the extension of matter were refuted. For it is one of the most important, and I believe the most certain, foundations of my physics . . ." (AT V 275; CSMK 364).[2] But Descartes's optimism about his account of the existence and the nature of matter does not find internal support in his own natural philosophy. As I shall argue below, one of the main problems with it has to do with the very nature of body: if extension, and extension alone, is the essential attribute of corporeal substance, then how can we distinguish between two bodies, since they are ultimately only extension? How do we know where one body ends and the other one begins?

This is the problem of individuation of physical bodies. It becomes a very important issue when moving from the general principles of physics to the description of particular cases where more bodies interact. In order to describe how the interactions take place, Descartes has to refer to the bodies involved—in other words, to identify each of the bodies in question. In addition to the internal problems of his theory, the individuation of bodies raises some problems in an apparently different area, namely, in his use of the illustrations in support of the text. As is well known, some of Descartes's books contain illustrations, which picture individual bodies that are "clearly and distinctly" separated from the other bodies. If the problem of his matter theory is to discriminate one body in respect of other bodies, in the case of the illustrations, the problem is to depict the bodies independently. Thus, the issue lies in the boundary we draw for each particular body.

With these observations in mind, solving the problem of the individuation of body becomes very urgent. Descartes claims he has solved the problem of the existence and nature of bodies in the second part of the *Principles*, but, as we shall see below, the internal tension existing in his theory was still a pressing issue for his followers. My aim in this chapter is to connect these two apparently distinct topics of Descartes's theory of matter and his use of illustrations in order to analyze the tension between the internal structure of matter theory and the drawings. For this, I shall start from a brief overview of Descartes's theory of matter and the problems raised by it, followed by a discussion of its reception in the works of two early Cartesians, Jacques du Roure and Gerauld (Géraud) de Cordemoy. At the end, I shall discuss how Descartes's illustrations complement the text in order to allow the reader to see individual bodies that are not distinctly individuated by his theoretical account.

DESCARTES'S MATTER THEORY

First, let us have a look at Descartes's theory of matter as it is presented in his later works, especially in the second part of the *Principles*. After the first part dedicated to metaphysics, Descartes turns to the discussion of physics, a discussion that is built on an argument for the existence of bodies. He starts from the observation that "everyone is quite convinced of the existence of material things" (AT IXb 63; CSM I 223) in order to reach his well-known conclusion: "there exists something extended in length, breadth and depth and possessing all the properties which we clearly perceive to belong to an extended thing. And it is this extended thing that we call 'body' or 'matter'" (AT IXb 64; CSM I 223).

This foundational step in the development of his physics is followed by a passage where the nature of bodies is explained. Descartes uses an eliminative argument, which starts from our perceptions that are further analyzed by the intellect. One by one, various attributes we perceive in a body are rejected due to their change over time. Similar to his famous example from the *Meditations*, where the piece of wax was changing its initial attributes and one by one, its smell, color, or shape either disappeared or was transformed (AT VII 30–31; CSM II 20), Descartes concludes in the *Principles* "[b]y the same reasoning it can be shown that weight, colour, and all other such qualities that are perceived by the senses as being in corporeal matter, can be removed from it, while the matter itself remains intact; it thus follows that its nature does not depend on any of these qualities" (AT IXb 65; CSM I 224). Ultimately, the attributes of corporeal substance are reduced to extension alone, which is considered to be the nature of body.[3]

At this point in his argument, Descartes acknowledges two difficulties for his theory:

> But there are still two possible reasons for doubting that the true nature of body consists solely in extension. The first is the widespread belief that many bodies can be rarefied and condensed in such a way that when rarefied they possess more extension than when condensed. . . . The second reason is that if we understand there to be nothing in a given place but extension in length, breadth and depth, we generally say not that there is a body there, but simply that there is a space, or even an empty space; and almost everyone is convinced that this amounts to nothing at all.
>
> (AT IXb 65–66; CSM I 225)

He addresses both of them, and so he continues with an account of rarefaction. Thus, rarefied bodies are defined as "those which have many gaps between their parts—gaps which are occupied by other bodies; and they become denser simply in virtue of the parts coming together and reducing or completely closing the gaps" (AT IXb 66; CSM I 225). A direct consequence of this view is that a body has a maximum density that cannot be surpassed; but for rarefaction this is not the case.[4] And here, Descartes uses the famous sponge example in order to explain that rarefaction is the product of the presence of a foreign body within the pores of the body in question (i.e. the sponge in this particular case). However, the explanation cannot stop here, because new questions will arise. A first issue is: how many pores can a body have? We know that the world is filled with bodies—Descartes will argue for this later on—and if we assume, on the one hand, that the number of pores a body has is indefinite, then one possible danger is to be unable to make any distinction at all between bodies. It might be possible to claim that the world is composed by one single body in a state of extreme rarefaction. On the other hand, if the number of pores is finite, it follows that a finite number of bodies will be allowed inside that body. Now, in connection with our problem, another question has to be addressed: how is it possible to distinguish between one body and other bodies that can fill in the pores of a particular body?

Descartes proceeds to the discussion of the "second reason for doubting," the possibility of void, which needs to be clarified. He recognizes that it is possible to perceive "empty spaces" as existing in the world, but, he adds, this does not mean that there are no bodies in them. Descartes differentiates between two meanings of "vacuum," a philosophical one ("that in which there is no substance whatsoever") and a common use ("usually refers not to a place or space in which there is absolutely nothing at all, but simply to a place in which there is none of the things that we think ought to be there" (AT IXb 71–72; CSM I 230)). Thus, even if we do not directly perceive that there are bodies, we can "arrive at knowledge of the shapes [sizes] and motions of particles that cannot be perceived by the senses" (AT IXb 321; CSM I 288). But let us leave aside, for the moment, the problem of the particles that are not directly perceived by our senses and return to Descartes's argument against the existence of void. He bases his rejection of void on the contrary nature of the concept: if void is understood as a space free of bodies, then it means that it should occupy a space. In this context, he discusses an example that will also be taken by his followers: a thought experiment that involves a vessel filled with air, which he assumes to be completely emptied by God.[5] The question is: what will happen in this case? Descartes's answer is very simple: the walls of the vessel will collide because there is no other body inside them. "Every distance is a mode of extension, and therefore cannot exist without an extended substance" (AT IXb 73; CSM I 230). The content of the vessel is not important, as it can be lead, water, air, or any kind of thing; what matters is rather that the corporeal substance will be of the same type. This is a further confirmation for Descartes that the volume describes the corporeal matter, not its weight or density (AT IXb 73; CSM I 231).

Having provided an explanation of the phenomena of rarefaction and condensation and after the rejection of void, Descartes reacts against atomism. He offers two arguments against the existence of atoms.

First, the argument of the Supreme Being: if we accept the existence of atoms, this means that we accept the existence of small bodies that cannot be further divided. But this implies that God cannot divide them, which of course is false as God is omnipotent (AT IXb 74; CSM I 231).[6] This reasoning looks more like an *ad hoc* argument because Descartes does not use a similar one when he discusses the problem of the created substances: why is it impossible for God, who is omnipotent, to make more than two separated substances?[7]

The second argument is a direct consequence of the way Descartes defines the nature of body.[8] Considering that bodies are defined by their extension, this means that they can be divided no matter how big or small they are, which further implies that it is impossible to stop the division in order to find an indivisible body. The differences between bodies seem to be based, first, on extension and, second, on the relation between a given body and its surroundings. The world is, thus, a collection of bodies, all having the same nature, or, as Descartes puts it: the world is a plenum composed by an "indefinite" extension (AT IXb 74; CSM I 232).[9] In order to justify

The Vanishing Nature of Body in Descartes's Natural Philosophy 15

the use of "indefinite," Descartes appeals to the imagination: we must conclude that the world is indefinitely extended because we can conceive something further than what we presume to be the limits of the world.

Something should be added about this second argument. When Descartes tried to support his view that the essence of body is extension, one of the main points was to make a clear demarcation from atomism. Now, he rejects atomism because it is not compatible with the consequences that can be derived from extension. Assuming that his argument does not imply any circularity in this respect, the problem of individuation still stands. If the division goes all the way down, then is a divided body *one* body or is it composed of *two* or *more* bodies? It will be difficult to differentiate between bodies and parts of bodies in a similar manner with what we noticed earlier about bodies occupying the "gaps" of a rarefied body.

Descartes seems to have a solution for this, for later on he defines "body" in the following way: "by 'one body' or 'one piece of matter' I mean whatever is transferred at a given time, even though this may in fact consist of many parts which have different motions relative to each other" (AT IXb 76; CSM I 233). But this creates a link between body and motion, the latter of which was earlier defined in the following terms: "motion is the transfer of one piece of matter, or one body, from the vicinity of the other bodies which are in immediate contact with it, and which are regarded as being at rest, to the vicinity of other bodies" (AT IXb 76; CSM I 233). This implies that a body will be considered in motion if its spatial relation to its surrounding bodies changes. But at the same time, this definition of motion will be applied to any other body, making it harder to say which body is moving. This can create some confusion, especially if we consider the body in respect to Descartes's explanation of divisibility:

> For what happens is an infinite, or indefinite, division of the various particles of matter; and the resulting subdivisions are so numerous that however small we make a particle in our thought, we always understand that it is in fact divided into other still smaller particles.... We cannot grasp in our thought how this indefinite division comes about, but we should not therefore doubt that it occurs. For we clearly perceive that it necessarily follows from what we <already> know most evidently of the nature of matter, and we perceive that it belongs to the class of things which are beyond the grasp of our finite minds.
> (AT IXb 82–83; CSM I 239)[10]

When applied to the illustrations, this theory will raise even more problems, something that we shall address here. First of all, Descartes's matter theory is problematic because of the last presented feature: the indefinite divisibility of matter makes it harder to speak about some particular body, because at every level bodies can be split apart. The second problem refers to the role of the observer. Descartes's theory is dependent on the observer, so that a body can be perceived as being in motion by someone while other observers consider it at rest. And finally, a third issue deals with the status of resting bodies: do they retain their individuality further, or do they become part of their surroundings?

In order to clarify these problems, we shall make use of one of Descartes's illustrations from the second part of the *Principles* (Figure 1.1).

Used by Descartes to exemplify how the motion of the bodies AB and CD in the surrounding medium EFHG takes place, this drawing is not at all "clear and distinct" but rather expresses all three difficulties. In the illustration he offers, it can be seen that the motion (either of the body AB or of the body CD) is considered in relation to the domain EFHG, which is at rest. If we take the case of two bodies AB and CD, they can be equally considered to move if one of them is taken to be at rest. But this is just the simple way to interpret it, and a more careful analysis will allow us to derive some further consequences concerning Descartes's theory as presented so far.

First of all, it is possible to raise the problem of the division of the body AB. As it is displayed in the illustration (Figure 1.1), it seems to be possible to divide it in two separate bodies, A and B. But, according to the previously discussed theory, Descartes's bodies are indefinitely divided, which means that the body A can always be split into A1, A2, and so on. Moreover, the division can go all the way down, because there are no atoms in Descartes's world, so each body will be (at least potentially) divisible.

A second difficulty comes from the way both "body" and "motion" are defined. Even if we do not raise for the moment the problem of circularity

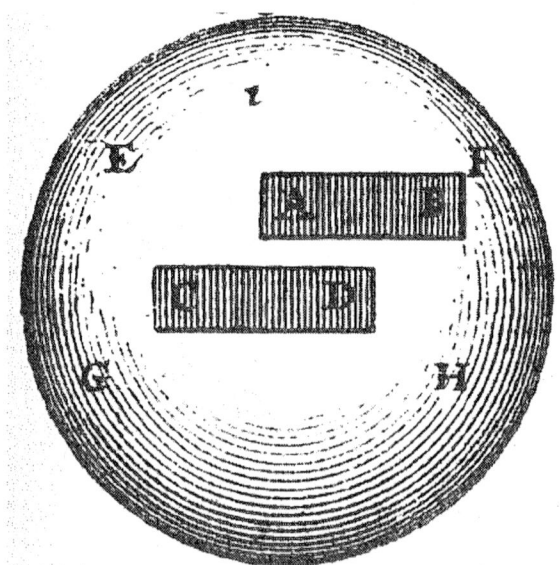

Figure 1.1 Diagram for *Principles* II, art. 30, AT IXb (rear).

The Vanishing Nature of Body in Descartes's Natural Philosophy   17

and keep to the first issue, we can still ask: why do we consider one body instead of two; in our particular example, why is it considered *the body* AB and not *the bodies* A and B moving together? Descartes's illustration can be taken in both ways and it seems that what is important here is the role of the observer. In Descartes's own terms, a body is "whatever is transferred at a given time," and this can vary from moment to moment, not to mention that it is relative to the observer and to the accuracy of one's perception. As our next example will show, this is mainly a problem of continuity between the relative rest of a body at one instant and its relative rest at another. Let us imagine that a group of people share the experience of the motion of body AB. At one time, a person goes out of the room, while everybody else witnesses the separation of body A from body B. After a couple of moments of independent motion, A and B will move again jointly, such that the body A will be in the same relation with the body B, the joined-up bodies changing their situation only in respect to the surroundings EFHG, exactly as in the beginning. At this point, the person who had left the room returns and sees the same situation as in the beginning. Now, the question is: how many bodies are moving from the point of view of that person? All others have seen that there are two distinct bodies, but this observation could have been accidental, because they could have gone out of the room together with their colleague and returned at the same time. Moreover, that motion not only depends on the observer but is also dependent on the accuracy of the perception. The last part is explained by the separation of the bodies A and B in our example, which means that the observer can wrongly take the bodies to be in reciprocal rest while they are moving very slowly one in respect to the other such that the motion is difficult to perceive.

Another observation concerning Descartes's example is about the situation of the body CD if it is taken to be at rest.[11] Compared to the body AB, which is in motion, the situation of CD seems to be clear, but what is the relation of CD with the surroundings EFHG? Descartes's starting assumption was that the surrounding medium, EFHG, was at rest. Now that the body CD is also at rest, the problem is how to differentiate it from the surroundings.

This can lead to some strange results, as can be seen in Figure 1.2.

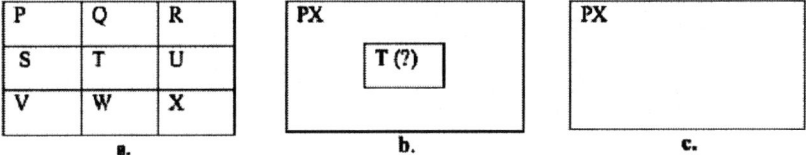

Figure 1.2   The dissolution of Cartesian bodies: The case of solid bodies.

Take nine similar, solid bodies P, Q, R, S, T, U, V, W, and X, and assume they are contiguous to one another (Figure 1.2a).[12] Further, assume them to be at rest one with respect to the others, such that, body T, for example, does not move with respect to its neighboring bodies.[13] In Descartes's matter theory, the problem with this is that it should be impossible to distinguish the body T from the medium PX (Figure 1.2b), as it can be observed in the last figure (Figure 1.2c). If T dissolves into the body PX because it is not moving in respect to its surroundings, then Descartes's physics does not seem to be able to give an acceptable explanation of rest.[14]

In her book on Descartes, Marjorie Grene stumbled upon the problem of the collections of bodies. Thus, Grene asks: "How many bodies are there? Minds are countable, since God creates each one at conception. But how many extended substances there are it is hard to say" (Grene 1985: 100). However, our analysis revealed that the problem is not to count bodies, but to distinguish them from the surrounding bodies, as I have shown in Figure. 1.2. In fact, Grene touches the main problem later on: "If we want to count bodies, they must be clearly and distinctly separable" (ibid.). As we can see from Figure 1.2c, the body T cannot be perceived distinctly. It can only be distinct in the imagination, which has no bearing on the way things are in the world. In imagination alone anyone can draw lines anywhere. Therefore, it follows that physics will be reduced to a geometry of imaginary shapes, which ultimately will make bodies vanish. Thus, Descartes's matter theory seems to lead to the dissolution of bodies. Although there are certain situations in which bodies dissolve themselves into a medium, some Cartesians like Gerauld de Cordemoy pointed out that the body T can also continue to maintain its individuality.

CARTESIANS FACING THIS PROBLEM

One of the first Cartesians, Gerauld de Cordemoy, is well known for his occasionalist solution and for his atomist theory of matter. Considered by many to deviate from Descartes's theory of matter, this original view is very important for our analysis.[15] In his own theory of matter, Cordemoy starts from the ordinary knowledge of the existence of both bodies and matter. He claims that the two concepts have been poorly understood in natural philosophy and that their careful examination is required for any system of philosophy, and especially for physics:

> We know that there are bodies and that their number is almost infinite. We also know that there is matter, but it seems to me that we do not have distinct enough notions of them, and that almost all of the errors of ordinary physics come from this.
>
> (Cordemoy 1666: 1)[16]

Cordemoy offers a simple solution to this problem by giving definitions for both terms, followed by five propositions in which their main attributes are presented. He conceives "bodies" as extended substances, while "matter" is not identical with these "bodies"—as it had been for Descartes—but is defined as a collection of bodies arranged in various ways: "The bodies are extended substances" and "The matter is a collection of bodies."[17] The main attributes of bodies are: figure, impenetrability, being in a place (*lieu*), motion, and rest. The crucial point lies in his definition of "impenetrability," as Cordemoy rejects the possible division of bodies on the basis of his new understanding of substance: "as each body is only a self-same substance, it cannot be divided; its figure cannot be changed; and it is so necessarily continuous that it excludes all the other bodies; this is called *impenetrability*."[18] This sentence offers both a new characterization of bodies and a metaphysical claim in favour of what Cordemoy calls a substance. While a body can change some of its attributes, Cordemoy finds that it is necessary to have at least one immutable criterion of individuation and distinction, and this is impenetrability. In other words, each body is a continuous, figured substance and hence impenetrable. Even if these properties represent the main characteristics of atoms, Cordemoy refers to "bodies" and does not use the term "atom" in his philosophy.

Cordemoy's metaphysical assumption is thus different from Descartes's. As Sophie Roux states: "Cordemoy marks the metaphysical base and the anti-Cartesian significance of his atomism: it is precisely because it is a substance that an atom is indivisible and distinct from its surroundings."[19]

If for Descartes the nature of body, or matter, is pure extension, for Cordemoy it has to be more than that. According to his view, a body is an indivisible *unity* characterized by impenetrability, a certain figure, and a spatial location, which is defined by its motion or rest. As stated earlier, impenetrability is the attribute that guarantees the indestructibility of a substance because it will not allow a given body to be divided into smaller ones. Thus, the *unity* of substance provides the metaphysical background of Cordemoy's atomism as he departs from Descartes precisely in this point, in the interpretation of divisibility.

And Cordemoy marks his departure by listing three objections against Descartes's philosophy.[20] His first objection deals with the term "indefinite," which, according to Cordemoy, is coextensive with "infinite." In his view, the use of "indefinite" cannot produce anything more than confusion. The second objection is more important and it can be split in two parts. One part has traditionally been identified as expressing the circularity in Descartes's definition of motion and the second part is concerned with the status of a resting body in the world.[21] Finally, Cordemoy's third objection to Descartes refers to the way in which various motions a body can have in its own parts are explained.

In what follows, I shall discuss only the second "inconvenience" of Descartes's matter theory. Cordemoy begins this critique with the following difficulty: "matter itself is an extended substance, which means that it is not possible to conceive of an independent body without assuming a motion."[22] This apparently small inconvenience encompasses an epistemological as well as a physical dimension. If matter is conceived as extension, following Descartes, it will be impossible to "know a body" (in the sense of having a clear and distinct perception of it) without making appeal to motion.[23] It follows that it is impossible to conceive a body at rest with respect to its surroundings. Put differently, a body will lose its distinctiveness once it stops moving in respect to its neighboring bodies. We can exemplify this with Descartes's own illustration, which we have seen earlier (Figure 1.1). Unless bodies AB and CD are conceived as moving, they will be indiscernible from the mass of the big body EFHG. Furthermore, is A a body in its own right, or is it just a component of the body AB?

In fact, Cordemoy recognises the problem very well:

> Thus, when we discuss a solid or some liquid which has parts that cannot be distinguished, we must examine what their effects are, then we must take into account which shapes are most likely to produce those effects; and we must believe that we have assumed well the shapes of the parts which compose a solid or a liquid, when we designated one that can explain all their effects.
>
> (Cordemoy 1666: 26)[24]

Now, this issue does not merely pertain to the particular illustration to which we have referred so far (Figure 1.1). Other examples can be found, and we can obtain a better understanding both of the consequences of Descartes's view as well as of the implications of Cordemoy's

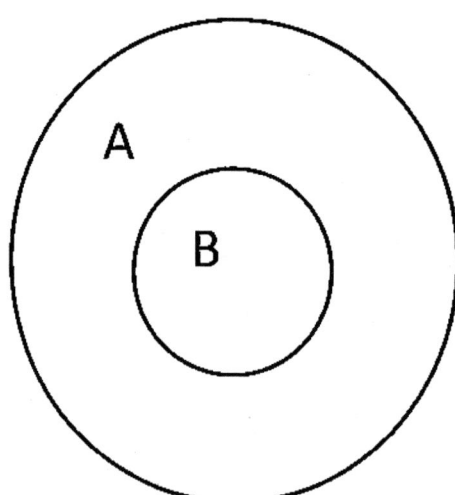

Figure 1.3 The dissolution of Cartesian bodies: The case of bodies carried around by other bodies.

The Vanishing Nature of Body in Descartes's Natural Philosophy

critique if we look at the next example, which is represented in Figure 1.3. Take body B as being surrounded by body A, and further, consider body B at rest with respect to the body A, regardless of previous or future relations.

This example is not hypothetical but can easily be applied to Descartes's explanation of the motion of the planets, which states that planets do not move, but are transported by the surrounding fluid of the vortex.

In this respect, Descartes's argument in the *Principles* was that:

> We must bear in mind what I said above about the nature of motion, namely that if we use the term "motion" in the strict sense and in accordance with the truth of things, then motion is simply the transfer of one body from the vicinity of the other bodies which are in immediate contact with it, and which are regarded as being at rest, to the vicinity of other bodies. . . . It follows from this that in the strict sense there is no motion occurring in the case of the earth or even the other planets, since they are not transferred from the vicinity of those parts of the heaven with which they are in immediate contact, in so far as these parts are considered as being at rest. Such a transfer would require them to move away from all these parts at the same time, which does not occur.
> (AT IXb 113–114; CSM I 252)

But if neither the earth nor the planets move, the question emerges as to how they maintain their distinctness and individuality. Descartes's answer lies in one of the premises of his argument: the heavens surrounding the planets have a fluid consistency. And one of the main features of any fluid body—as Descartes argued in the second book of the *Principles*—lies in the behavior of its constituent particles.[25] These do not keep the same relations with one another but are in constant motion and very agitated, which allows for a notable difference between their motion and the one of the planets they are transporting.

However, Descartes's argument raises at once a new problem. The boundary interaction between the heavens and the planet is described in the following terms:

> Since the celestial material is fluid, at any given time different groups of particles move away from the planet with which they are in contact, by a motion which should be attributed solely to the particles, not to the planet. In the same way, the partial transfers of water and air which occur on the surface of the earth are not normally attributed to the earth itself, but to the parts of water and air which are transferred.
> (AT IXb 114; CSM I 252)

In proceeding from the visible world to the small bodies that compose it and, in particular, to the interactions between all the small particles that can be found at the boundary between heavens and planets, it is impossible to establish a determinate boundary. Moreover, according to Descartes's "proper" definition of motion, these particles will have a different motion than the planets and the surrounding heavens, forcing us to reduce the question for the case of smaller and smaller bodies, *ad indefinitum*. How can we then differentiate between planets and the heavens, since the boundary between them is indeterminate?

This leads us to one problem discussed by another Cartesian, Jacques du Roure. In his *La Physique expliquée suivant le sentiment des anciens et nouveaux Philosophes; & principalement Descartes* of 1653, du Roure argued that the atoms give rise to problems of individuation, so he rejected their existence.[26] However, later on, in his *Abrégé de la vraye philosophie, lequel en contient avéque les six parties & leurs tables: les définitions, les divisions, les sentences et les questions principales* (1665), du Roure adds a very strange passage that solves Descartes's problem, but makes his previous position look contradictory. Du Roure claims that when the extremities of two bodies touch, "The parts of the touching bodies, which are in relative rest one to the other, [these parts] are naturally indivisible. What would be the reason to divide them in one way and not in the other?"[27] He continues:

> I give these parts, the name of simple bodies, or physical and natural atoms: in order not to confuse them with these other atoms, which are so-called corporeal, but which nevertheless do not have parts or figures, which do not occupy space, and finally which are something that one does not conceive what they are.
>
> (Du Roure 1665: 116)[28]

Du Roure's attempt to solve one of the problems raised by Descartes's theory of matter comes at a cost, namely, the addition of atomic "simple bodies" to the realm of physics. Now, these simple bodies are said to be different from the "other atoms," which are "something that we don't know what they really are." But while the rejection of atomism is not a strange philosophical position for a Cartesian, du Roure's particular explanation, which replaces the atomist account with atomic *simple bodies*, is obviously an utterly un-Cartesian claim. His difficulty in finding a solution for this problem is obvious, as he tries to dissociate his view from the atoms of the atomists. But the description given for "ces autres Atomes," the "Atomes corporels," is completely implausible. Whose atoms are they? Certainly not Democritus's, Epicurus's, Gassendi's, not even the "bodies" of Cordemoy.

DESCARTES'S ILLUSTRATIONS, OR FROM THE VISIBLE TO THE INVISIBLE

In the second discourse of his *Les Météores*, Descartes gives an account of the production of vapors based on the interactions between small (and invisible) corpuscles of water. As we can see in Figure 1.4, despite the fact

The Vanishing Nature of Body in Descartes's Natural Philosophy 23

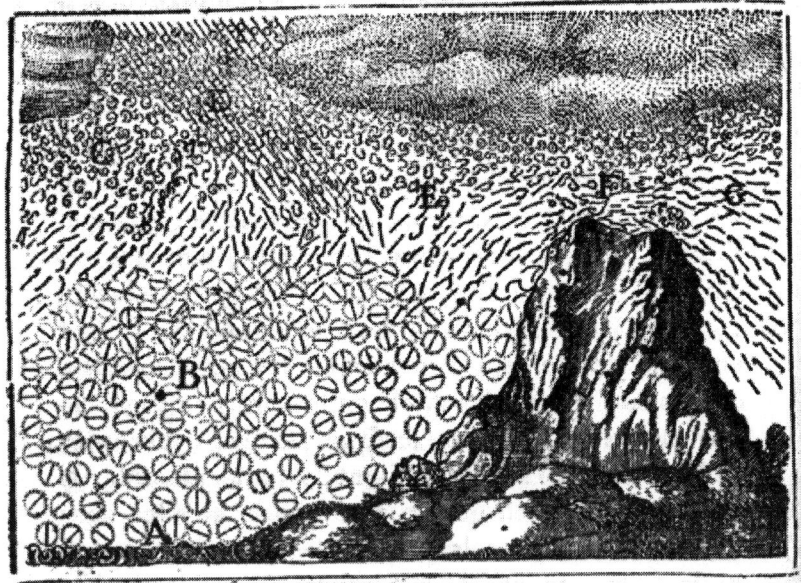

Figure 1.4 Les Météores: AT VI 242.

that Descartes did not know the form or the size of these corpuscles, he chose to depict them in a particular way.

Later on, in one of his epistolary exchanges, when challenged to give a more detailed explanation of this subject, Descartes not only backed off, but seems to admit the possibility of accepting other forms and figures of the small bodies:

> 11. Of course I do not claim to be certain that the particles of water are shaped like certain animals, but only that they are elongated, smooth and flexible. Now, if we can find any other shape that enables us to explain all their properties just as well, I shall be perfectly willing to adopt that instead; but if no other can be found, I do not see what difficulty there could be in imagining them specifically to have that shape, seeing that they must necessarily have some shape or other, and the one I suggested is particularly simple.
>
> (AT II 43; CSMK 101)

In this letter addressed to Reneri for Pollot (April or May 1638), Descartes follows the path from visible to invisible in order to depict something that cannot be perceived through the senses, but, when he is asked to justify his position, he admits a lesser degree of certainty for the knowledge of the watery particles. However, the problem becomes even more complex if we put it into the context of the theory of matter presented so far. We know that a body is individuated according to its motion in respect to the surrounding bodies. This illustration shows us a heap of individual bodies that have the same form (B), and even if we accept Descartes's transduction from the visible to the invisible, the problem of the individuality of these bodies will still be at issue.[29] Descartes justifies his position by claiming that: "[n]o one who uses his reason will, I think, deny the advantage of using what happens in large bodies, as perceived by our senses, as a model for our ideas about what happens in tiny bodies which elude our senses merely because of their small size" (AT IXb 319; CSM I 287).[30]

But this is not only a matter of comparing what happens at the visible level with the explanation of the invisible phenomena.[31] As pointed out earlier, Descartes provides a complex explanation for the way in which we reach the knowledge of these small, invisible bodies. Thus, in the fourth part of the *Principles*, he argues:

> First of all I considered in general all the clear and distinct notions which our understanding can contain with regard to material things. And I found no others except for the notions we have of shapes, sizes and motions, and the rules in accordance with which these three things can be modified by each other—rules which are the principles of geometry and mechanics. And I judged as a result that all the knowledge which men have of the natural world must necessarily be derived from these notions; for all the other notions we have of things that can be perceived by the senses are confused and obscure, and so cannot serve to give us knowledge of anything outside ourselves, but may even stand in the way of such knowledge. Next I took the simplest and best known principles, knowledge of which is naturally implanted in our minds; and working from these I considered, in general terms, firstly, what are the principal differences which can exist between the sizes, shapes and positions of bodies which are imperceptible by the senses merely because of their small size, and, secondly, what observable effects would result from their various interactions. Later on, when I observed just such effects in objects that can be perceived by the senses, I judged that they in fact arose from just such an interaction of bodies that cannot be perceived—especially since it seemed impossible to think up any other explanation for them.
> (AT IXb 321; CSM I 288)

Descartes notoriously claims that explanation is made on the basis of three notions that can be applied to bodies, namely shape, size, and motion. They are announced to be taken in respect to other similar notions and especially in respect to "the simplest and best known principles." This process should also be doubled by an analysis of the perceived effects, which is said that "in fact arose from just such an interaction of bodies that cannot be perceived." From here it follows that, while it is impossible that they be perceived and directly tested, Descartes's judgments in respect to the small invisible parts of bodies have to be true because they are derived from

The Vanishing Nature of Body in Descartes's Natural Philosophy 25

certain first principles, which are exactly the principles we have discussed so far. If we are going to derive the properties of the invisible small parts of bodies from the matter theory presented so far, we will face the same difficulties as we did in the earlier discussion of the theory. And one of the main consequences of this kind of approach is that bodies will not be able to retain individuality as far as Descartes's tensions between the definitions of "body" and "motion" (with the addition of rest) will come into play.

Thus, depicting small, individual bodies can be a good rhetorical tool to provide models for how things work in the world, but still, they cannot provide certainty.[32] And this is not due to a wrong comparison between small and large bodies, but mainly due to the "first principles" that explain what a body is and how it interacts with the other bodies. Avoiding the use of illustrations is one possible solution, because it will avoid the tension between theory and images, but a more fundamental step is required in order to solve the problem.

CONCLUDING REMARKS

What we are left with in Descartes is just a world that has to be described "as if," in a similar way with Garber's conclusion for the problem of individuation:

> I shall continue to talk as if Descartes is dealing with a world of individual bodies, colliding with one another, at motion and at rest with respect to one another. But, in the end, I suspect that this is something that he is not entitled to, and this is something that, if true, would seriously undermine his whole program.
>
> (Garber 1992: 181)

Descartes concedes that his natural philosophy is only one way that the inner workings of the world might be. For instance, when he refers to the way in which the world was produced by God, he claims:

> Just as the same craftsman could make two clocks which tell the time equally well and look completely alike from the outside but have completely different assemblies of wheels inside, so the supreme craftsman of the real world could have produced all that we see in several different ways.
>
> (AT IXb 327; CSM I 289)

And this line of reasoning was further developed by other Cartesians. For example, in the late seventeenth century, the Oratorian philosopher Bernard Lamy published a book called *Entretiens sur les sciences* (1683), in which he raised a similar issue:

> To have the right to imagine that one understands things, one must be able to explain them as one would explain a watch which one opens so that one sees the movement and shape of its parts . . . [Descartes] tries to explain the whole world and its effects like a watchmaker who wishes to understand the way in which a watch shows the hours. . . . Just as one discovers with the help of the telescope those objects whose distance hides them from our eyes, so likewise one sees things whose small size makes them unobservable without the aid of a microscope. That is what needs to be done in order to philosophize. Because everything which appears in the body is just like the case of the watch which hides the mechanism. It is therefore necessary to open this case; however, in nature the springs are so small that our eyes cannot observe their subtlety without assistance. . . . One must recognize however that in a great many things, even with the aid of the microscope, pneumatic machines and chemistry, we still cannot penetrate what Nature had decided to conceal from us. We do not see what is inside. What can a physician do, therefore, except conjecture?
> (Lamy 1966: 256, 257–258, 259)[33]

We have the same analogy with the clocks and the knowledge of their hidden mechanism. But Lamy moves from the comparison between visible and invisible to the role of imagination and ultimately to the claim that we need conjectures in physics. His solution comes in the form of a question, but still it points out to a hypothetical approach in physics. Thus, some parts of physics can still be considered "certain" while others fail the test. Physics does not have to be accepted or rejected as a whole, because even if the matter theory has to be better founded, it can be later improved.

It follows that when Descartes wrote to More about his matter theory as "one of the most important, (. . .) the most certain, foundation of my physics," and claimed that all the physics is based upon it, he failed to see the problem of the individuation of bodies in particular and the problem of the foundation of physics in the matter theory in general. A careful analysis of Descartes's arguments in physics showed us that bodies tend to "vanish" in his plenum. And the problem of his matter theory does not only consist in the way in which bodies are conceived in respect to motion—something that has already been pointed out by a number of scholars—but another more serious problem can be found in the way he explains rest. Extended substance, then, will be perceived clearly, but will lack the required distinctness to identify (and to illustrate) a particular body. From the Cartesians we have discussed here, Du Roure merely recognizes this problem of Descartes's philosophy, but he does not engage in finding a solution. Cordemoy digs much deeper but at the cost of the entire Cartesian matter theory. His atomist solution comes to replace Descartes's account of matter. And he succeeds as long as the individuation crisis finds a solution in the concept of atom. Moreover, even the vapors of water, depicted by Descartes in Figure 1.4 in the region B, can be saved as individuals by an atomist theory.[34]

NOTES

1. See also Gaukroger 2006: 321: "The system that Descartes proposes in the *Principles* requires a balance of mechanics and matter theory which was to act as a model for many natural philosophers throughout the seventeenth and eighteenth centuries, offering a good fit between mechanics and matter theory."
2. English translations of Descartes's writings are taken, where available, from CSM and CSMK.
3. For a more extensive discussion of this reductive argument, see Garber 1992: 77–80.
4. Comparing Descartes's view with that of the Aristotelians, Des Chene (1996: 351) states that: "For the Aristotelians, rarefaction and condensation were changes in the intensive quantity associated with rarity and density. The standard case is the change of one element into another—the same quantity of water, when transmuted by heat into air, is said to increase tenfold volume. . . . Descartes's response is to show how the appearance of rarefaction or condensation can be saved without admitting any quantity other than quantity of extension, and while maintaining the conservation of matter." After this, Des Chene refers to other seventeenth-century sources where a similar explanation can be found; notably to Sebastien Basso and Isaac Beeckman.
5. Another famous version of this thought experiment involves an "empty" room, where the walls simply collide such that there will not be any room. See Rohault 1987: 27.
6. Usually Descartes does not use such arguments in his physics, because the absolute power of God cannot be understood by the finite human mind. Still, Descartes argues that even if we do not have a complete knowledge of God, we are still able to understand and to speak about His attributes. See Descartes's letter to Mersenne from January 21, 1641 (AT III 284; CSMK 169). For a larger discussion of the philosophical context of the debates concerning the absolute and the ordained power of God, see Osler 1994.
7. Descartes seems to emphasize *potentia dei absoluta* here, while he avoids using such an argument in other places of his natural philosophy. For instance, in the context of the former discussion about void, why is it impossible for God to take out the whole content of the vessel and not to maintain it as such?
8. There are some disagreements in the secondary literature concerning this point. For an atomist reading of Descartes, see Roux 2000.
9. For the use of the words "infinite" and "indefinite," see McGuire 1983.
10. This is a passage where Descartes discusses the possibility of motion "en un cercle imparfait, et le plus irrégulier qu'on saurait imaginer" (AT IXb 81). He refers to the division of "some parts of matter," as the French version of the text shows: "à savoir une division de quelques parties de la matière jusqu'à l'infini, ou bien une division indéfinie, et qui se fait en tant de parties, que nous n'en saurions déterminer de la pensée aucune si petite, que nous ne concevions qu'elle est divisée en effet en d'autres plus petites. (. . .) Et bien que nous n'entendions pas comment se fait cette division indéfinie, nous ne devons point douter qu'elle ne se fasse, parce que nous apercevons qu'elle suit nécessairement de la nature de la matière, dont nous avons déjà une connaissance très distincte, et que nous apercevons aussi que cette vérité est du nombre de celles que nous ne saurions comprendre, à cause que notre pensée est finie" (AT IXb 82–83). If in this context the explanation was given for the bodies that were reaching a particular place when they travelled on the circles, the reasoning can still be applied to the entire extended substance,

because if every body moves on these circles described here by Descartes, at one point they will have to pass through the places where this division takes place. It follows that all matter is "indefinitely" divisible. The notion of divisibility proposed by Descartes at this point of his argument was criticized early on in the seventeenth century. Many authors, like Newton in "De Gravitatione" (Newton 2004: 23–25), or Cordemoy 1666, or Leibniz, noticed that the indefinite divisibility is not a clear concept, and they objected that Descartes uses a vague concept for what should be infinite.

11. Garber noticed the problem of rest in Descartes's theory of matter: "Since bodies are individuated through motion, where there is no motion, there can be no distinct individuals. That is, two bodies that we are inclined to say are at rest with respect to one another must, for Descartes, really be two parts of a *single* body. And so, it would seem, Descartes is committed to the rather paradoxical view that all individual bodies must be in motion," Garber 1992: 178.
12. The previous state of the bodies is not an issue for this example.
13. The example can work for any of the bodies depicted. The following example is not a problem for the body T, but for any contiguous two similar bodies, taken to be at rest one in respect to the other.
14. This would not have been a problem, as far as his plenum implies that all bodies are in motion. However, Descartes does use rest when he describes motion. A particular body is said to be in motion only because its neighbouring bodies are taken to be at rest.

Even more, in my analysis this is an important issue in respect to the role of the illustrations. How can Descartes depict a body at rest, since it should lose its individuality if we are going to stick to his theory strictly?
15. For a general presentation of Cordemoy's philosophy, see Ablondi 2005.
16. Cordemoy 1666: 1: "On sçait qu'il y a des Corps & que le nombre en est Presque infiny: On sçait aussi qu'il y a de la matiere; mais il me semble que l'on n'en a pas de notions assez distinctes, & que c'est de là que viennent Presque toutes les erreurs de la Physique ordinaire."
17. Cordemoy 1666: 2, 3: "Les Corps sont des substances estenduës" and "La Matiere est un assemblage de corps."
18. Cordemoy 1666: 2: "Comme chaque corps n'est qu'une mesme substance, il ne peut estre divisé; sa figure ne peut changer; & il est si necessairement continu qu'il exclud tout autre corps; ce qui s'appelle *impenetrabilité*."
19. This follows Garber 1992. See Roux 2000: 265: "Cordemoy marque le fondement métaphysique et la signification anti-cartésienne de son atomisme: c'est en tant qu'il est substance qu'un atome est indivisible et distinct de tous ceux qui l'entourent."
20. For a larger discussion of Cordemoy's critique, see Ablondi 2005: 30–43.
21. For the circularity problem, see Ablondi 2005: 36 and Garber 1992: 175–181. Basically, the accusation of circularity refers to the way Descartes defines both body and motion. As we have seen above, each of them is defined in respect to the other, making the understanding of body problematic.
22. Cordemoy 1666 : 11–12: "la matiere mesme est une substance estenduë, c'est qu'ils ne sçauroient faire concevoir un corps à part, sans supposer un mouvement."
23. See also Garber 1992: 178–179.
24. Cordemoy 1666: 26: "Ainsi quand on propose une masse ou quelque liqueur dont les parties ne se peuvent discerner, on doit examiner quels en sont les effets, ensuitte l'on doit considerer quelles figures sont les plus propres à produire de tells effets; & l'on doit croire que l'on a bien supposé la figure des parties, qui composent une masse, ou une liqueur, quand on en assigne une, qui peut rendre raison de tous leurs effets."

25. See AT IXb 94; CSM I 245: "fluids are bodies made up of numerous tiny particles which are agitated by a variety of mutually distinct motions; while hard bodies are those whose particles are all at rest relative to each other." For Descartes's statement that the heavens are fluid, see AT IXb 112; CSM I 251.
26. This is a very Cartesian book, where du Roure claims in a number of places that he follows Descartes's philosophy. *La Physique* is a part of a larger textbook, *La Philosophie divisée en toutes ses parties, établie sur des principes évidents, et expliquée en tables et pur discours, tirés des anciens et des nouveaux auteurs et principalement des Péripatéticiens et de Descartes*, which was published in 1654.
27. See du Roure 1665: 116: "Les parties des corps qui se touchent, & qui reposent l'une contre l'autre, sont naturellement indivisibles. Car pourquoy seroient-elle divisées en un endroit, plûtôt qu'en un autre?"
28. Ibid.: "Je donne à ces parties, le nom de Corps simples, ou encore d'Atomes Physiques & naturels: afin de ne les confondre pas aveq ces autres Atomes, qui sont à ce qu'on dit corporels, mais qui neanmoins n'ont point de parties ni de figures, qui n'occupent point d'espace, enfin qui sont ce qu'on ne conçoit pas qu'ils soient."
29. For transduction, see Mandelbaum 1964: 63.
30. A similar comparison between the visible and the invisible is made also by some of the Cartesians. For instance, Cordemoy made a connection between what happens at the visible level and how small bodies behave at the invisible level: "S'il est vray que le moindre corps doit avoir figure & peut ester meu; & s'il est vray enfin que les loix de la nature soient les mesmes à proportion pour les petites & pour les grandes masses, ou peut raisonner de la figure & des mouvemens des corps que l'on ne void pas, par ce que l'on connoît des figures & des mouvemens des masses que l'on apperçoit," Cordemoy 1666: 66.
31. For the notion of comparison in Descartes, see Galison 1984.
32. For a general discussion of Descartes's use of illustrations, see Lüthy 2006.
33. As quoted in Desmond Clarke 1989: 175.
34. An early version of this paper was presented in the colloquium "Vanishing bodies and the birth of modern physics" (Bucharest, June 30–July 2, 2008), and I thank the audience for its reception. Thanks are also due to my colleagues from *The Center for the History of Philosophy and Science* (Radboud University Nijmegen), who had the patience to discuss a preliminary version of this article. Not least, I would like to express my gratitude to the two editors of this volume for their careful reading and helpful comments on previous drafts.

BIBLIOGRAPHY

Ablondi, F. (2005) *Gerauld de Cordemoy: Atomist, Occasionalist, Cartesian*, Milwaukee: Marquette University Press.
Barber, K. F. and Garcia, J. J. E., eds. (1994) *Individuation and Identity in Early Modern Philosophy: Descartes to Kant*, Albany, NY: State University of New York Press.
Clarke, D. (1989) *Occult Powers and Hypotheses: Cartesian Natural Philosophy under Louis XIV*, Oxford: Oxford University Press.
Cordemoy, G. de (1666) *Le discernement du corps et de l'âme en six discours: pour servir à l'éclaircissement de la physique*, Paris.
Descartes, R. (1664) *L'homme de René Descartes et la formation du fœtus avec les remarques de Louis de La Forge* (with a preface by Claude Clerselier), Paris.

——. (1996) *Œuvres de Descartes*, revised edn, 11 vols, eds. C. Adam and P. Tannery, Paris: Vrin.
Des Chene, D. (1996) *Physiologia: Natural Philosophy in Late Aristotelian and Cartesian Thought*, Ithaca: Cornell University Press.
Du Roure, J. (1653) *La physique expliquée suivant le sentiment des anciens et nouveaux philosophes: et principalement de Descartes*, Paris.
——. (1654) *La Philosophie divisée en toutes ses parties, établie sur des principes évidents, et expliquée en tables et pur discours, tirés des anciens et des nouveaux auteurs et principalement des Péripatéticiens et de Descartes*, Paris.
——. (1665) *Abrégé de la vraye philosophie, lequel en contient avéque les six parties & leurs tables: les définitions, les divisions, les sentences et les questions principales*, Paris.
Festa, E. and Gatto, R. (2000) *Atomismo e continuo nel 17. secolo: atti del Convegno internazionale Atomisme et continuum au 17. siècle, Napoli, 28–29–30 aprile 1997*, Naples: Vivarium.
Galison, P. (1984) "Descartes's comparisons: from the invisible to the visible," *Isis*, 75: 311–326.
Garber, D. (1992) *Descartes' Metaphysical Physics*, Chicago: University of Chicago Press.
Gaukroger, S. W. (2003) *Descartes' System of Natural Philosophy*, Cambridge: Cambridge University Press.
——. (2006) *The Emergence of a Scientific Culture: Science and the Shaping of Modernity, 1210–1685*, Oxford: Clarendon Press.
Grene, M. (1985) *Descartes*, Minneapolis: University of Minnesota Press.
Grosholz, E. (1994) "Descartes and the individuation of physical objects" in eds. K. F. Barber and J. J. E. Garcia 1994, pp. 41–58.
Kusukawa S. and Maclean, I., eds. (2006) *Transmitting Knowledge: Words, Images, and Instruments in Early Modern Europe*, Oxford: Oxford University Press.
Lamy, B. (1966) *Entretiens sur les sciences*, Paris: Presses Universitaires de France.
Lüthy, C. (2006) "Where logical necessity becomes visual persuasion: Descartes's clear and distinct illustrations" in eds. S. Kusukawa and I. Maclean 2006, pp. 97–133.
Mandelbaum, M. (1964) *Philosophy, Science, and Sense Perception: Historical and Critical Studies*, Baltimore: Johns Hopkins University Press.
McGuire, J. E. (1983) "Geometrical objects and infinity: Newton and Descartes on extension" in ed. W. Shea 1983, pp. 69–112.
Newton, Sir I. (2004) *Isaac Newton: Philosophical Writings*, ed. A. Janiak, Cambridge: Cambridge University Press.
Osler, M. (1994) *Divine Will and the Mechanical Philosophy*, Cambridge: Cambridge University Press.
Rohault, J. (1987) *System of Natural Philosophy, Illustrated with Dr. Samuel Clarke's Notes, Taken Mostly out of Sir Isaac Newton's Philosophy*, New York: Garland Publishing.
Roux, S. (2000) "Descartes atomiste?" in eds. E. Festa and R. Gatto 2000, pp. 211–273.
Shea, W. ed. (1983) *Nature Mathematized: Historical and Philosophical Case Studies in Classical Modern Natural Philosophy*, Dordrecht: Reidel.

2 The New Matter Theory and Its Epistemology
Descartes (and Late Scholastics) on Hypotheses and Moral Certainty

Roger Ariew

As Descartes fully understands, the new matter theory explains the behavior of sensible bodies by reference to imperceptible particles. So the question arises, how can we arrive at the knowledge of the shapes, sizes, and motions of these particles? The answer involves the epistemic status of hypotheses, but the role of hypotheses in Descartes's philosophy is not clear, or it seems to have undergone some change, and the Cartesians do not seem to have accepted Descartes's view fully. Now, it has already been pointed out that Descartes was not as *a prioristic* about scientific method and the use of hypotheses as is usually thought, or at least that he became less so in his later years (see, for example, Blake, Ducasse, and Madden 1960; Laudan 1981; Garber 1978, and 2001b; and Clarke 1982), and that the Cartesians, while maintaining Descartes's propensity for mechanistic explanations, became more empirical and pursued aggressively a quasi-hypothetical-deductive method (Clarke 1989; McClaughlin 2000; also Ariew 2006). But the motivations for these shifts are not clear: it is not useful to treat Descartes and the Cartesians as sleepwalkers, darkly perceiving the hypothetico-deductive nature of science, as has sometimes been done. Science may or may not have a single method; hypothetico-deductivism may or may not be that method. Even if it were, this in itself could not explain why the Cartesians accepted some form of it, if they did. So I wish to investigate the various uses Descartes and the Cartesians made of hypotheses and the reasons they gave for those uses. In this chapter I limit my discussion to Descartes and some of his immediate predecessors. As will become obvious, a fair portion of my discussion concerns Descartes's notion of moral certainty, which Descartes uses to distinguish between nonhypothetical first principles about general things and hypothetical ones about particular things. Descartes's usual view was that his hypotheses could be grounded in nonhypothetical, self-evident principles, that he had or could provide such a derivation. By Part IV of the *Principles of Philosophy*, he knew that such a demonstration would be futile. Descartes's opinion in the *Principles* is that his hypothetical principles are not absolutely but merely morally certain, meaning that there is at least some logical connection and coherence in them, such that his physics would have to be rejected and taken only as a fiction, or else it all has to be accepted and not be rejected until another is found more capable of explaining all the phenomena of nature. The key concept for Descartes is thus "moral certainty," a term he consciously borrows from the late scholastics. I try to make sense of Descartes's concept by reference to contemporary scholastics texts, namely those of Roderigo Arriaga, Eustachius a Sancto Paulo, and Francisco Suárez. However, because of how Descartes distinguishes between absolute and moral certainty, to draw the full picture, I would have to discuss as well the status of the method of doubt in the second half of the seventeenth century. In brief, my overall argument would have been that the rejection of the method of doubt, which is the underpinning of moral certainty, causes many Cartesians no longer to distinguish between the absolutely and the merely morally certain—between that which we cannot doubt and that about which we have no doubt although we could doubt it—and thus to treat all principles on a par with one another. But I leave this extension of my inquiry to another occasion.

There are significant discussions of the status of hypotheses in Descartes's *Principles*, parts III and IV. There, Descartes indicates that he is aware of the long tradition of hypotheses in astronomy and relates his own use of hypotheses to it.[1] On the basis of the relativity of motion at the phenomenal level, Descartes simply claims that the question about whether the earth is in motion cannot be fully resolved by the appearances. As a result, astronomers have invented three different hypotheses to account for the phenomena: those of Ptolemy, Copernicus, and Tycho (*Principles* III, art. 15, AT VIIIa). Descartes then rejects Ptolemy's hypothesis as not adequate to account for all observations and in particular for the recently observed phases of Venus (*Principles* III, art. 16, AT VIIIa). Descartes asserts that the choice between the other two hypotheses is underdetermined by the appearances (*Principles* III, art. 17, AT VIIIa). He gives preference to Copernicus's hypothesis over Tycho's on grounds of "simplicity" and on the fact that Tycho "has not sufficiently considered the true nature of motion" (*Principles* III, art. 17–18, AT VIIIa). Descartes proceeds to prefer his own (pseudo-Copernican) hypothesis on the same grounds: he is proposing the hypothesis that seems to him "the simplest, most true, and most useful for knowing the phenomenon as well as for enquiring into natural causes" (*Principles* III, art. 19, AT VIIIa). He specifically warns his reader he is not claiming that his "hypothesis should be received as entirely in conformity with the truth, but only as a hypothesis <or supposition that could be false>" (*Principles* III, art. 19, AT VIIIa and IXb for the bracketed phrase).

Later on, in *Principles* III, Descartes considers more generally all hypotheses, not just those given in astronomy. He argues that it is not likely that the causes from which all the phenomena can be deduced are false: as long as he "has used evident principles, deduced the results by mathematical reasoning, and accounted exactly for all experience, it would be injurious to God to believe false the causes of the natural effects discovered in this way" (*Principles* III, art. 43, AT VIIIa). Still, Descartes does not want to assert that all the hypotheses proposed by him are true. He prefers people to take what he has written "as a hypothesis that could be quite distant from the truth" (*Principles* III, art. 44, AT VIIIa). However, he indicates that if what is deduced from his hypothesis is in conformity with experience, then the hypothesis "would be no less useful to life than if it were true, <because we could use it in the same way to produce the desired effects>." (*Principles* III, art. 44, AT VIIIa and IXb for the bracketed phrase). He even asserts that he will assume some hypotheses he believes to be false; he allows himself to imagine some simple and intelligible principles that are contrary to the account of creation from Genesis, which he takes to be true (*Principles* III, art. 45, AT VIIIa). Descartes thinks that all the bodies in the universe are composed of the same matter, a matter divisible into parts that are variously moved in circular motions, and that there is always an equal quantity of motion in the world. But he cannot determine how large are the parts into which the matter is divided, nor with what speed they move, nor what circles they describe. "These things could have been ordered by God in an infinity of ways and it is only by experience <and not the power of reason> that one can know which way he has chosen from all of these." That is why Descartes believes he is free to assume whatever he wishes about the division of the parts and their motion, as long as what he deduces from his hypothesis "agrees <entirely> with experience." And he proceeds to assume that God at first divided the matter into equal parts, etc., something he knows contradicts Genesis, but which is a simpler and more intelligible supposition (*Principles* III, art. 46, AT VIIIa and IXb for the bracketed phrase). Descartes's whole doctrine can be summarized by his comment to the Jesuit Denis Mesland in a 1645 letter about the *Principles*:

> I dare say that you would find at least some logical connection and coherence in it, such that everything contained in the last two parts [that is, *Principles* III and IV] would have to be rejected and taken only as a pure hypothesis or even as a fable, or else it all has to be accepted. And even if it were taken only as a hypothesis, as I have proposed, nevertheless it seems to me that, until another is found more capable of explaining all the phenomena of nature, it should not be rejected.
> (*To Mesland*, May 1645, AT IV 216–217)

Yet prior to the 1640s, Descartes had another view. Descartes discusses the status of hypotheses in the 1637 *Discourse on Method* and in two of its appended essays, the *Dioptrics* and *Meteors*. At the end of the *Discourse*, referring to the hypotheses he used at the beginning of the two essays, he says that people should not be shocked by the fact that he calls some things "suppositions" and that he does not seem to want to prove them (*Discourse* VI, AT VI 76: *suppositions* in the *Discourse*; *hypotheses* in the Latin translation of the *Discourse*, called *Specimina philosophiae*: AT VI: 582). He asserts, "I only called the things suppositions so that it can be known that I think I can deduce them from the first truths I have previously explained, but that I wished expressly not to do it"—and this because he did not wish to reveal everything at once (*Discourse* VI, AT VI 76). It is the main difference between the accounts of the *Discourse* and of the *Principles*. In the *Principles*, Descartes would have not promised a derivation of his hypothetical principles from his first truths but would have argued only for the coherence of the whole lot. And, indeed, in the *Meteors*, Descartes uses principles he claims have not been sufficiently explained *as yet*; he calls them "suppositions," declares that he will be able to render them so extremely simple and easy that it will not be difficult to believe them, and refers to the use he has made of such suppositions about light in the *Dioptrics* (*Meteors* I, AT VI 233: *suppositions*; *hypotheses* in the *Specimina*, AT VI: 652). There he divulges that he will not be talking about the *nature* of light, but will be using some comparisons that "would help in conceiving it in the easiest way . . . imitating in all this the astronomers whose suppositions are almost all false or uncertain . . . and yet allowing one to derive consequences from them that are true and certain" (*Dioptrics* I, AT VI 83: *suppositions*; *hypotheses* in the *Specimina*, AT VI: 585).

In a series of letters from 1638, the astrologer Jean-Baptiste Morin objected to Descartes's treatment of hypotheses. He agreed that the appearances of the celestial movements can be derived from the supposition of the earth's stability as well as from the supposition of its mobility and claimed that experience is not sufficient to prove which of the two causes is the true one. But he challenged Descartes, asserting that if his suppositions are not any better than those of the astronomers he is imitating, he is going to do no better than them—perhaps even worse: "there is nothing easier than to adjust some cause to an effect" (*Morin to Descartes*, February 22, 1638, AT I 539). Descartes responded to Morin in the same way he responded to others then. In a letter to a Jesuit, Descartes insisted that he was speaking hypothetically about light in the *Dioptrics* precisely because he had explained the matter "in a most ample and most curious manner" in the treatise he called *On Light* and did not want to repeat himself, but only wanted to represent an idea by means of comparisons and "shadows" (*To Vatier*, February 22, 1638, AT I 562). This was, of course, an exaggeration by Descartes, because the treatise in question, now known as *The World*, in which he claimed to have explained the matter amply, was itself a treatise that proceeded hypothetically. Descartes even referred to the treatise as a fable in a letter to Mersenne of 1630 (*To Mersenne*, November 25, 1630, AT I 179), and in the work itself, as it has come down to us, Descartes says that he will shorten his account by using a fable, in which he hopes the truth will still come through sufficiently (*The World* V, AT XI 31), and he states that he does not promise to give exact demonstrations of everything, that he will draw a picture with shadows as well as with clear colors, having no intent other than to relate a fable (*The World* VII, AT XI 48). As late as 1638, responding to Mersenne about whether what he has written about refraction is a demonstration, Descartes answers that he believes it to be so, "at least to the extent that it is possible to give demonstrations in that matter without having demonstrated the principles of physics by metaphysics" (*To Mersenne*, May 27, 1638, AT II 141), and he adds, parenthetically, that this is what he hopes to do one day, but it is something that has not yet been done. To Morin, Descartes replies by agreeing with what he asserted, but claiming that "the effects I explain have no other causes than the ones from which I deduce them, even though I reserve to myself the right to demonstrate this in another place" (*To Morin*, July 13, 1638, AT II 200).

The difference between Descartes's position *circa* 1637 and his position *circa* 1644 is that by 1644 Descartes has given up the possibility of deriving all his principles with the same kind of self-evidence and certainty.[2] He still thinks that some of his principles are certain and not mere hypotheses, but accepts others as inevitably hypothetical. In 1637, he was claiming that he could ground the hypothetical principles in nonhypothetical, self-evident ones, that he had or could provide such a derivation. By 1644, he knew that such a demonstration would be futile. The argument does not seem to change in other respects. There are obviously differences in the historical situations between the two periods. In 1637, in the aftermath of Galileo's condemnation, Descartes was publishing only the portion of his physics that he thought would not be controversial and burying the rest of it; in 1644, after having issued the *Meditations*, Descartes was confident enough to publish all of his physics, including its metaphysical foundations. But Descartes was at his most publicly confident self in the earlier period, where he was promising more, though claiming that the more was given in a work he was not publishing.[3]

At the end of *Principles*, Part IV, Descartes reflects generally on the method he uses in his physics and compares it with that of others. First he claims that he has not used any principle not accepted by all the philosophers of every age, including Aristotle. He has "considered the shapes, motions and sizes of bodies and examined the necessary results of their mutual interaction in accordance with the laws of mechanics, which are confirmed by reliable everyday experience" (*Principles* IV, art. 200, AT VIIIa). According to Descartes, no one has doubted that "bodies move and have various sizes and shapes, that their various different motions correspond to these differences in size and shape; and that when bodies collide bigger bodies are divided into many smaller ones and change their shapes." The difference between his approach and that of others is that he considers that "there are many particles in each body which are so small that they are not perceived with any of our senses" (*Principles* IV, art. 201,

AT VIIIa). Descartes then deals with the question of the nature of these unsensed bodies. Among other things, he rejects the Democritean "suppositions" that the minute bodies are indivisible and that there are voids around them, calling these suppositions inconsistent. But, he reiterates, "no one can doubt that there are in fact many such particles" (*Principles* IV, art. 202, AT VIIIa). He declares that he leaves it to others to judge whether his suppositions have been consistent. In the French edition of the *Principles*, he also adds that he leaves it to others to decide whether the results that can be deduced from his suppositions have been "sufficiently fertile"; he asserts that he rejects "all of Democritus's suppositions," with the one exception of "the consideration of shapes, sizes, and motions," and rejects "practically all the suppositions of other philosophers" as well (*Principles* IV, art. 202, AT IXb).

In the 1647 French edition of the *Principles*, Descartes describes the method he has used with respect to his suppositions. He has first considered in general all the clear and distinct notions the understanding can contain with regard to material things—those of shapes, sizes, and motions—and the rules in accordance with which these three things can be modified by each other—that is, the principles of geometry and mechanics. So he has concluded that all the knowledge people have of the natural world must be derived from these notions. Next, he has deduced the principal differences between the bodies that are imperceptible by the senses merely because of their small size and the observable effects that would result from their various interactions. Then, when he has observed just such effects as perceived by the senses, he has concluded that they in fact arose from such an interaction of bodies that cannot be perceived—"especially since it seemed impossible to think up any other explanation for them" (*Principles* IV, art. 203, AT IXb). His legitimation for this seemingly abductive procedure is an analogy: people who are experienced in dealing with machinery like a clock "can take a particular machine whose function they know and, by looking at some of its parts, easily form a conjecture about the design of the other parts, which they cannot see" (*Principles* IV, art. 203, AT IXb).

Descartes extends his analogy about such machines as clocks to make clear the limitations of the explanations of phenomena referring to corpuscles our senses do not perceive. Two clocks identical on the outside may indicate the time equally well but use different operating mechanisms. So also God could have produced the phenomena we perceive in innumerably different ways. As a result, the causes postulated by Descartes to explain some effects may correspond to the phenomena manifested by nature, but may not be the ones by which God produced those effects: "With regard to the things that cannot be perceived by the senses, it is enough to explain their possible nature, even though their actual nature may be different" (*Principles* IV, art. 204, AT VIIIa and IXb). These explanations, according to Descartes, are only *morally* certain, that is, they suffice for the conduct of life, although, given the absolute power of God, they *can* be doubted. In the French edition of the *Principles*, Descartes adds: "Thus those who have never been in Rome have no doubt that it is a town in Italy, even though it could be the case that everyone who has told them this has been deceiving them" (*Principles* IV, art. 205, AT VIIIa and IXb). In this way, Descartes distinguishes between two kinds of certainty, one he calls moral (and perhaps physical) and another he calls absolute (or mathematical). The situation is different with absolute certainty, which, according to Descartes, we possess for mathematical demonstrations, the knowledge that material things exist, and the evidence of all clear reasoning that is carried on about them: "Absolute certainty arises when we believe that it is wholly impossible that something should be otherwise than we judge it to be" (*Principles* IV, art. 206, AT IXb). So absolute certainty accrues to metaphysical principles that have passed the test of hyperbolic doubt and to the general physical principles that can be derived from them. Moral certainty accrues to the physical principles about particular things that cannot be perceived. We do not have real doubts about these principles, but they fail the test of hyperbolic doubt, because we understand that God could have brought about things in some other way.

Descartes uses another example to illustrate moral certainty. He refers to a code-breaker who has decoded a message and who is certain of his solution, but who understands that another solution might be possible. He states: "It is true that his knowledge is based merely on a conjecture, and it is conceivable that the writer . . . encoded quite a different message; but this possibility is so unlikely that it does not seem credible" (*Principles* IV, art. 205, AT VIIIa and IXb). Descartes adds in the French edition: "especially if the message contains many words." He concludes, cashing in his analogy:

> Now if people look at all the many properties relating to magnetism, fire, and the fabric of the entire world, which I have deduced in this book from just a few principles, then, even if they think that my assumption of these principles was arbitrary and groundless, they will still perhaps acknowledge that it would hardly have been possible for so many items to fit into a coherent pattern if the original principles had been false.
> (*Principles* IV, art. 205, AT VIIIa and IXb)

Still, Descartes argues that, at bottom, his explanations "possess more than moral certainty," because at least the most general results have been deduced in an unbroken chain from the first and simplest principles of human knowledge (*Principles* IV, art. 206, AT VIIIa and IXb). Descartes's promise to provide a derivation of his principles from self-evident ones remains, but it is now limited to the general principles given in Parts I and II; those of Parts III and IV about the nature of particular things are now irremediably hypothetical, that is, just morally certain as opposed to absolutely certain.

There has been a fair amount of commentary about moral certainty, but I think the concept is still not well understood. I wish to provide evidence for three claims about moral certainty in Descartes: 1) Descartes frequently used the concept before his formal definition of it in the *Principles*—he did so even in 1637–1638, when he was claiming to Morin that the effects he explained had no causes other than the ones from which he deduced them—thus, to call something morally certain was a commonplace in Descartes's vocabulary before 1644 (as it was after). 2) Descartes borrowed the concept from the schoolmen, for whom it was also a commonplace—though, of course, he made it his own (this is not to deny that there could be some evolution, as well as some nonformal uses of the concept in Descartes). And 3) against most commentators, despite what could be inferred from Descartes's examples of code-breaking and of knowing where Rome is, moral certainty should not be equated with high probability.

An important instance in which Descartes claims to have reached moral, not absolute, certainty is the case of animals and machines, described in the 1637 *Discourse*. We have moral certainty that an entity using language or acting through knowledge, "imitating our actions as closely as morally possible" will be human, and not animal or machine: "for it is morally impossible for there to be enough different organs in a machine to make it act in all the contingencies of life in the same way as our reason makes us act" (*Discourse* V, AT VI 57). In case one is wondering whether these passages should be understood in some way other than Descartes's asserting that we have moral certainty of these possibilities or impossibilities, there is another passage in the *Discourse* where Descartes makes clear use of moral certainty in the standard way. He argues:

> If there are still people who are not sufficiently convinced of the existence of God and of their soul by the arguments I have proposed, I would have them know that everything else of which they may think themselves more sure—such as their having a body, there being stars and an earth, and the like—is less certain. For although we have a moral certainty about these things,[4] so that it seems we cannot doubt them without being extravagant, nevertheless when it is a question of metaphysical certainty, we cannot reasonably deny that there are adequate grounds for not being entirely sure about them.
>
> (*Discourse* IV, AT VI 37–38)

Descartes's claim that we have only moral certainty about there being stars and an earth does not seem altogether consistent with his claim that we can know that the effects he explains about such things as the stars and the earth can have no other causes than the ones from which he deduces them. But still, that is what his promise of a deduction would seem to indicate.

There are other interesting instances of moral certainty in Descartes's correspondence from the period before 1644–1645. In very much the same fashion as with the *Discourse* question about the kind of certainty we have that an animal cannot perfectly imitate our actions, Descartes claims that a machine cannot fly like a bird: "A machine that can be sustained in the air like a bird can be constructed *metaphysically speaking*—for, according to me, birds are such machines—but not *physically* or *morally speaking*, because there would have to be springs so subtle and so strong collectively that they could not be fashioned by men" (*To Mersenne*, August 30, 1640, AT III 163–164). He also refers to an experiment as "morally impossible" (*To Mersenne*, March 11, 1640, AT III 40); he says that it is "morally impossible" to remove all the printers' errors from his manuscript (*To Mersenne*, July 22, 1641, AT III 415–416); and he rejects the objection that according to his philosophy we would have no certainty that a priest is holding a host at the altar, for, Descartes says, "Who has ever said, even among scholastic philosophers, that there was more than moral certainty about such things?" (*To Mersenne*, April 21, 1641, AT III 359).

In his published work, in a passage of the *Sixth Set of Replies*, referring to the creation of the eternal truths, Descartes talks about a king being "the efficient cause of a law, although the law itself is not a thing that has physical existence, but is merely what they call a moral entity."[5] "Moral" here has nothing to do with morality or ethics, but indicates that the entity is less than real in the same way that moral certainty is a grade of certainty less than absolute. Moreover, in the *Seventh Set of Replies*, Descartes defends himself against the charge that in the *Meditations* he claimed some knowledge when he said he knew that there was no danger in his renouncing his beliefs: "in that passage I was merely speaking of knowing in the moral way which suffices for the conduct of life. I frequently stressed that there is a very great difference between this type of knowledge and the metaphysical knowledge that we are dealing with here" (*Seventh Set of Objections and Replies*, AT VII 475).

Descartes continues using the concept after 1644–1645, with some interesting variations. He refers to the result of an experiment as "something that cannot happen morally" (*To Mersenne*, April 26, 1643, AT III 653). He talks about "physical and moral causes, which are particular and limited, as opposed to a universal and indeterminate cause" (*To Mesland*, May 2, 1644, AT IV 111), refers to reasons as "being neither mathematical, nor physical, but only moral" (*To Huygens*, November 30, 1646, AT IV 788), and to knowing that we have "examined matters as far as we are morally able" (*To Christina*, November 20, 1647, AT V 83–84). In an extremely important letter to Mesland, Descartes asserts that "when a very evident reason pulls us to one side, even though, morally speaking, we cannot choose the opposite side, absolutely speaking, however, we can."[6]

Two of the instances of the use of the qualifier "moral"—Descartes saying that we have a moral not metaphysical certainty about there being stars and an earth and his referring to a moral entity as opposed a physical one—are translated in different places with some added phrases, the first, "*ut loquntur Philosophi,"* in the *Specimina* (again, this is the 1644 Latin translation of the *Discourse*), and the second, *"comme ils disent en l'École,"* in the 1647 French translation of the *Sixth Replies*. In these cases, the translators (Étienne de Courcelles for the *Specimina* and Claude Clerselier for the *Objections and Replies*) are indicating what seems obvious to them, that the phrase has been borrowed from scholastic terminology. That does seem fairly clear; in fact, the distinction becomes so commonplace that it gets codified in the first edition of the *Dictionnaire de L'Académie française*, though considerably weakened so that moral certainty becomes "apparent certainty" and seems allied with probability: "One says 'moral certainty' so as to say verisimilitude or apparent certainty; and then 'moral' is in opposition to 'physical.' Thus one says, 'You are not given physical certainty, but there is some moral certainty in this.' . . . One says 'morally speaking,' so as to say "verisimilously and according to all the appearances.'"[7]

There is an interesting passage in which a three-fold distinction is made among moral, physical, and absolute certainty in Roderigo Arriaga's 1632 *Cursus philosophicus*:

> Certainty is three-fold, moral, physical, and metaphysical. Moral certainty is what we have when our reasons are indeed fallible physically, though infallible morally speaking, i.e., almost infallible, as, for example, the certainty I have about the existence of Naples, from what has been said by so many knowledgeable and honest men who assert it and make me certain that Naples exists, although, because it is not physically impossible that they should all lie, I am not physically certain of this existence. . . . Physical certainty is what rests on physical principles which cannot, in accordance with the nature of the thing, be otherwise, as, e.g., the certainty I have about Peter's running, which I see; for the thing really can be otherwise, at least by [God's] absolute power, insofar as God can miraculously make it appear to me that Peter is running, even though he is not really running; and he can do the same in other matters; therefore that certainty is not called metaphysical and supreme, but natural or physical. Finally, metaphysical certainty is that by which the object is presented in such a way that in relation to every power it cannot be otherwise, as the certainty I have about God's existence, or about such principles as *Each thing either is or is not*, or *Things which are the same as a third thing are the same as each other*, and the like, or about all the mysteries revealed by God, which cannot be false, even in relation to God's absolute power.
>
> (Arriaga 1632, Logica, disp. 16, sec. 4: 226 col. a)[8]

Moral certainty is defined and illustrated, in Arriaga's three-fold distinction, in the same way as Descartes defines and illustrates it—though Descartes uses Rome, not Naples, as his example of moral certainty. Descartes also plays with a three-fold distinction—he talks of reasons being neither mathematical nor physical but only moral—though ultimately he defines just moral and absolute certainty. He can probably make room for physical certainty in the fashion of Arriaga—something that rests on physical principles that cannot be otherwise without invoking God's absolute power—but he would not think that seeing Peter running is certain in that way. Thus, physical certainty for Descartes would seem to collapse into moral certainty or no certainty at all. Arriaga's distinction is also interesting because of where it is given in his philosophical course. It occurs in Part I, *Logic*, just after the discussion of syllogism, in an examination of demonstration and the certainty provided by it, together with a discussion of the differences among demonstrative science, opinion, and faith (Arriaga 1632, Logica, disp. 16, sec. IX: 238–239). Arriaga talks about three degrees of certainty in demonstrative science but carefully distinguishes the certainty we obtain in science from the opinion we derive from probable reasoning.

Given Descartes's distaste for scholastic logic, it seems unlikely that he would have read these passages directly, but likely that he would have been acquainted with the general views represented. A book Descartes is known to have read, Eustachius a Sancto Paulo's *Summa Philosophiae Quadripartita*, does not provide a discussion of the kind of certainty we obtain through demonstration, but does provide one about the differences among science, opinion, and faith, with clear implications for there being different kinds of certainty:

> We may ask how demonstrative science is related to opinion and faith. The three are alike in being capable of truth . . . but with this difference, namely, that science is a state that is always true, while opinion and faith can be false. Then again, science and opinion rely on reason—science on necessary reason, opinion on probable reason—while faith relies on authority alone. Faith is a condition that has two aspects, one directly implanted, and called divine faith, and the other acquired, and called human faith. The former is always certain like science, because it depends on divine authority, which can never deceive or be deceived; but the latter can be doubtful, like opinion, because it depends on human authority, which can both deceive and be deceived. . . . Now the three conditions are distinguished in terms of their objects, insofar as science relates to something that is necessary in virtue of its cause, while opinion relates to something that is probable by reliable signs, and finally faith relates to something testified by authority. And hence the three are distinguished in virtue of their corresponding acts: by science we know, by opinion we conjecture, and by faith we believe.
> (Eustachius a Sancto Paulo 1629, Logica, part III, tract 3, disc 1, q 4: 152)

Eustachius continues by arguing that it may happen that we have science, opinion, and faith with respect to the same thing, although the means we employ are different: it is science if the thing is demonstrated by necessary reason, opinion if it is inferred merely by probable reasoning, and faith if it is believed because of the authority of the testifier. Hence these three conditions can all be present in the same intellect with respect to the same thing. Since science and faith can both produce certainty, certainty comes in different kinds.

For Arriaga as for Eustachius, science differs from opinion in that the former is certain while the latter is merely probable. And, similarly as well, for Arriaga faith is given either immediately and supernaturally or mediately and naturally. It can be merely probable, or it can be certain, that is, morally, not metaphysically certain, in the same way that one knows where Rome is (Arriaga 1632, Logica, disp. 16, sec. IX: 239, col. a, b. Arriaga changes his illustration from knowing that Naples exists to knowing that Rome exists). This discussion about the difference between science, opinion, and faith is common in scholastic textbooks, as is the discussion of the kinds of certainty science and faith can provide. The Jesuits of Coimbra also propose three kinds of certainty: what they call certainty of the object, of the known, and of the knower ("Certitudo est triplex, objecti, cognitionis, et cognoscentis," Conimbricenses 1606: 696, col. b). The first is the usual demonstrative certainty about necessary things and the third has to do with the firmness of the intellect in adhering to the thing as a truth, while the second makes room for the certainty of faith. This is also the background for Francisco Suárez's discussion of the knowledge of the existence of God being based mostly on faith, though accompanied with practical and moral certainty, that is, what he calls *moral* self-evidence:

> ... this general notion is based on the tradition of the majority and is passed on from parents to children, from the more learned to the less. As a result, the general belief that God exists has grown and become accepted among all peoples. Hence this knowledge seems in large part to be due to faith, especially among the masses, rather than to the self-evidence of the matter; but it still seems to have been attended with practical and moral self-evidence, which is sufficient to oblige people both to assent to the truth of God's existence and also to propagate it. And accordingly we may easily understand everything that the Doctors of the Church say about knowledge of God being naturally implanted. (Suárez 1998, disp. 29, sec. 3, no. 35–37, vol. 2: 60.)

Suárez uses the concept of moral self-evidence or certainty without any fanfare, assuming the kinds of discussion common to the commentaries of the Conimbricenses, Eustachius, and Arriaga. In these discussions, moral certainty is a species of certainty, carefully distinguished from opinion and high probability.

The same can be said for Descartes in his own way. It is tempting to think that moral certainty is high probability because of the examples of the code-breaker who decodes a message and the person who is told about Rome. Are we not more secure in our decoding, given that we have broken a larger code than a smaller one? Is it not relevant that we are told about Rome from many sources as opposed to a few? Still, moral certainty and high probability are usually distinguished, as they are in the discussions of Arriaga or Eustachius. And despite Descartes's examples, his moral certainty does not admit of any degree. The case for our being morally certain that we could not construct a machine that flies like a bird might look to us like a case of high probability, but for Descartes it is, like his other cases, something beyond the pale. Building such a machine would be so difficult that Descartes is morally certain that it could not be done by us, though God has done it; the same for constructing a machine that actually uses language. Moral certainty suffices for the conduct of life, but not in the sense that it is a good rule of thumb or something highly probable. As with the scholastics, it is genuine certainty within its own sphere. If something is morally certain, we lack any reason to doubt it, though we could doubt it if we considered God's absolute power. Descartes's two kinds of certainty are thus dependent on our being able to construct hyperbolic reasons for doubt: Absolute certainty, the certainty attaching to his metaphysical principles and the principles about general things he deduces from them, passes that criterion, whereas moral certainty, the certainty attaching to physical principles about particular things, fails it.

Once one understands Descartes's peculiar notion of moral certainty and the role it plays in his system, it is easy to see what can become of it in the hands of followers who might reject metaphysics altogether, that is, the grounding of physical principles in metaphysical ones, or some aspects of the method of doubt. The rejection of hyperbolic doubt caused Cartesians no longer to distinguish between the absolutely and the merely morally certain—between that which we cannot doubt and that about which we have no doubt although we could doubt it—and thus to treat all principles on a par with one another. As a result, Cartesians became more empirical and pursued aggressively a (limited or quasi) hypothetical-deductive method. Of course, not all Cartesians followed the same path in their espousal of probabilism and a hypothetico-deductive method in physics. Here is typical paragraph supporting a hypothetico-deductive method ending up with high probability, not absolute or moral certainty. It is from the Preface to the second edition of Christiaan Huygens's *Traité de la Lumière* (1690):

> One finds in this work these kinds of demonstrations that do not produce as great a certainty as those of Geometry, and that even differ much from geometrical demonstrations, given that geometers prove their propositions

by certain and incontestable principles, while here principles are verified by conclusions derivable from them; the nature of these things does not allow any other treatment. It is always possible, however, to attain in this way a degree of probability (*vraisemblance*), which very often is little short of complete evidence. This is the case when things demonstrated by these assumed principles correspond perfectly to the phenomena that experiment has brought under observation—especially when there are a great number of them, and further, principally, when one can devise and predict new phenomena that should follow from the hypotheses one uses, and one finds that the effect corresponds to our expectations. But if all these proofs of probability (*vraisemblance*) are encountered in what I propose to treat, as it seems to me they are, this should be a very strong confirmation of the success of my inquiry, and it is scarcely possible that the facts are not just about as I represent them.

Huygens, who is not an orthodox Cartesian, is here following a path taken by closer followers of Descartes, from the heterodox Henricus Regius to the relatively more orthodox followers, Jacques Rohault and Pierre-Sylvain Régis (see Mouy 1934).

NOTES

1. In French, this is *hypothèses* and *suppositions* as nouns, with *supposer* as a verb; in Latin, these are *hypotheses*, *positiones*, and *ponere*.
2. Various commentators view Descartes's development differently. After all, there are sufficient texts available to be able to argue for the hypothetical structure of Descartes's science from his earliest writings. Even in the *Discourse*, Descartes states: "First, I have tried to find in general the principles or first causes of all that is or can be in the world, without considering anything but God alone, who created the world, and without deriving these principles from any other source but from certain seeds of truths that are naturally in our souls. After that I examined what were the first and most ordinary effects that could be deduced from these causes.... After this, passing my mind again over all the objects that have ever presented themselves to my senses, I dare say I did not notice anything in them that I could not explain easily enough by means of the principles I had found. *But I must also admit that the power of nature is so ample and so vast, and these principles are so simple and so general, that I notice hardly any particular effect without at once knowing that it can be deduced in many different ways from them*," AT VI 64–65 (emphasis mine). Cf. Garber 1978 and Gaukroger 1995.
3. Cf. Garber 2001b for more on this background and the relationship to the project of the *Rules* of his promise to derive the principles of physics from metaphysics.
4. The *Specimina* adds, "ut loquntur Philosophi [as the philosophers say]," AT VI 561.
5. *Sixth Replies* AT VII 436. French translation: "la volonté du roi peut être dite la cause efficiente de la loi, bien que la loi même ne soit pas un être naturel, mais seulement (comme ils disent en l'École) un être moral" (AT 9-1: 236).

6. To Mesland, February 9, 1645, AT IV 173, AT III 379, also gives a French version of this letter, To Mersenne, 27 May 1641: "*Morallement* parlant, il soit difficile que nous puissions faire le contraire, parlant neantmoins *Absolument*, nous le pouvons."
7. "On dit, Asseurance morale, seureté morale, pour dire, Asseurance vray-semblable, seureté apparente: & alors Morale s'oppose à Physique. Ainsi on dit, On ne peut pas vous donner des seuretez physiques, mais il y a une seureté morale à cela. . . . On dit, Moralement parlant, pour dire Vraysemblablement, & selon toutes les apparences," Moeurs, *Dictionnaire de L'Académie française* 1694.
8. Cited by Curley 1993: 16–17. Compare with Etienne Chauvin's *Lexicon philosophicum*, post-dating Descartes: "An act of the intellect is said to be morally certain when it assents to a truth that, although it can happen otherwise, is nevertheless so constant that doubt about it is contrary to good principles of action. . . . One is said to be physically certain when one assents to an object firmly, on account of an immutable principle in nature, or . . . to a truth that, although it is possible to think otherwise, is most constant, as long as the order of nature remains the same. . . . One is said to be metaphysically certain when one is assenting firmly to an object which is presented to me in such a way that it cannot be otherwise even by the absolute power of God, or when I assent to a truth which cannot even by thought be otherwise," cited by Curley 1993: 16, from Gilson 1913: 334.

BIBLIOGRAPHY

Ariew, R. (2006) "Cartesian Empiricism," *Revue roumaine de philosophie*, 50: 71–85.
Arriaga, R. (1632) *Cursus philosophicus*, Antwerp.
Blake, R. M., Ducasse, C. J., and Madden, E. H. (1960) *Theories of Scientific Method; The Renaissance through the Nineteenth Century*, Seattle: University of Washington Press.
Chauvin, E. (1692) *Lexicon philosophicum*, Rotterdam.
Clarke, D. M. (1982) *Descartes' Philosophy of Science*, University Park, PA: Pennsylvania State University Press.
———. (1989) *Occult Powers and Hypotheses: Cartesian Natural Philosophy under Louis XIV*, Oxford: Clarendon Press.
Conimbricenses. (1606) *Commentarii in universam dialecticam Aristotelis*, Coimbra.
Curley, E. (1993) "Certainty: psychological, moral, and metaphysical" in ed. S. Voss 1993, pp. 11–30.
Descartes, R. (1996) *Œuvres de Descartes*, revised edn, 11 vols, eds. C. Adam and P. Tannery, Paris: Vrin.
Dictionnaire de L'Académie française (1st edn, 1694).
Eustachius a Sancto Paulo (1629) *Summa philosophiae quadripartita*, Paris.
Garber, D. (2001a) *Descartes Embodied*, Cambridge: Cambridge University Press.
———. (2001b) "Descartes on knowledge and certainty: from the *Discours* to the *Principia*" in Garber 2001a, pp. 111–129.
Garber, D. (1978) "Science and certainty in Descartes" in ed. M. Hooker 1978, pp. 114–151.
Gaukroger S. W. (1995) *Descartes: An Intellectual Biography*, Oxford: Oxford University Press.
Gaukroger, S. W., Schuster, J. A., and Sutton, J., eds. (2000) *Descartes' Natural Philosophy*, London: Routledge.
Gilson, E. (1913) *Index scolastico-cartésien*, Paris: Félix Alcan.

Hooker, M., ed. (1978) *Descartes: Critical and Interpretive Essays*, Baltimore: Johns Hopkins University Press.

Huygens, C. (1690) *Traité de la Lumière*, Paris.

Laudan, L. (1981) "The clock metaphor and hypotheses: the impact of Descartes on English methodological thought, 1650–1670" in *Science and Hypothesis: Historical Essays on Scientific Methodology*, Dordrecht: Reidel.

McClaughlin, T. (2000) "Descartes, experiments, and a first generation Cartesian, Jacques Rouhault" in eds. S.W. Gaukroger, J. Schuster, and J. Sutton 2000, pp. 330–346.

Mouy, P. (1934) *Le développement de la physique cartésienne, 1646–1712*, Paris: Vrin.

Suárez F. (1998) *Disputationes Metaphysicae*, Hildesheim: Olms.

Voss, S., ed. (1993) *Essays on the Philosophy and Science of René Descartes*, Oxford: Oxford University Press.

Part II
Matter, Mechanism, and Medicine

3 Post-Cartesian Atomism
The Case of François Bernier

Vlad Alexandrescu

François Bernier (1620–1688) is one of the philosophers who did the most to enrich and disseminate Pierre Gassendi's philosophy. *L'Abrégé de la philosophie de M. Gassendi*, a seven-volume work published in French, went through three successive versions in 1674, 1678, and 1684, and soon became the main instrument whereby Gassendi's ideas were made familiar to the *honnête homme* of the late seventeenth century (Koyré 1980: 321). The aim of this chapter is to assess Bernier's place in the seventeenth-century debate between the Cartesians and the Gassendists, particularly in terms of his views on bodies and on the structure of matter.

A discussion of the *Abrégé* in its entirety lies beyond the scope of this chapter; such a discussion was the object of a special issue of the journal *Corpus* in 1992, at the time of the publication of a reprint of this work by Fayard (Bernier 1992). It has been clear, ever since the colloquium held on the occasion of the tercentenary celebration of Gassendi's death (Gassendi 1955), that more work needed to be done in order to clarify the position of Bernier's *Abrégé* relative to Gassendi's *Syntagma philosophicum* (published posthumously in 1658 by a group of his disciples led by Habert de Montmor) or to other texts by Gassendi, such as the *Animadversiones in librum X. Diogenes Laertis*. Alexandre Koyré and Bernard Rochot raised questions as to Bernier's fidelity to the text of the *Syntagma*, and proposed the hypothesis of a deviating Bernier.

However, the pursuit of these questions cannot gain momentum in the absence of a modern translation of the *Syntagma philosophicum*, a monumental and dense work, copiously amassing quotes and opinions drawn from the humanist school, of which Gassendi seems to have been the last representative in France. Certainly, Bernier undertook a heroic task in completely redistributing the contents of Gassendi's work according to a plan of his own devising, in cutting out many historical introductions, in trimming off a whole host of examples, and in omitting many encumbering doxographic elaborations. The *Abrégé* has the appearance of a French garden; it is written in a robust French language, although the Latin neologisms do resurface sometimes in Bernier's reshuffled philosophical terminology.

The question that has been formulated since 1955 is whether, besides this work of rearrangement and refashioning of Gassendi's doctrines in the *Syntagma*, Bernier may have developed his own philosophy, which would have altered the transmission of Gassendi's ideas to the French intellectual circles. Sylvia Murr, editor of Bernier's work, undertook a first evaluation of this supposed deviation, which she proposed was due to two distinct factors: on the one hand, what she called Bernier's "opportunism," or his "concern to integrate philosophy with the movement of ideas at the end of the century, so that he could more easily gain acceptance for Gassendi's philosophy," and on the other, a "difference in genius" between the two figures (Murr 1992: 118). This first investigation went no further than her 1992 article, although Sylvia Murr refined her approach in a subsequent article (Murr 1997b), where she also offered a comparative table of the contents of the *Syntagma* and of the *Abrégé*.

In his article in the same journal, Roger Ariew worked with a different set of methods and presuppositions, and reached several interesting conclusions. In the 1684 edition entitled "seconde édition, revue et augmentée par l'auteur" [second edition, revised and enlarged by the author], Bernier appended to the second volume of the *Abrégé* a section of *Doutes* on Gassendi's doctrine, which he had previously published in a separate volume in 1682. Analyzing the strategy of this text, Ariew concluded that Bernier's aim was to attack Descartes's physics as dogmatic in order to salvage the substance of Gassendi's physics, which he should have condemned for the same reasons (Ariew 1992). Ariew also gave a general evaluation of the Gassendists' strategies aimed at tempering public criticism in the intellectual climate of the second half of the seventeenth century and at protecting a philosophy that was at least as novel as the Cartesian, and that, in effect, prevented it from being placed on the Index.

In what follows, we will consider two chapters of the first edition of the *Abrégé* (1674), where Bernier argues in favor of the atomist hypothesis. We will compare this text with the equivalent chapters in Gassendi's *Syntagma*. We will then consider the differences and ask the question of Bernier's fidelity to Gassendi, without ignoring the fact that the *Abrégé* came out nineteen years after the latter's death. We will also consider what becomes of these two chapters in the enlarged and revised edition of the *Abrégé* (1684) in order to understand the development of Bernier's Gassendism in confrontation with the various intellectual tendencies of the time. Finally, we will look at the *Doutes*, where Bernier significantly distances himself from some of Gassendi's postulates, and draws a comparative table of Gassendist and Cartesian tenets, as a means to define and justify a position of his own.

The chapters in Bernier we will look at, "De l'existence des atomes" [Of the existence of atoms] (Bernier 1674: 28–51) and "De la fluidité, de la solidité, de l'humidité et de la sécheresse" [Of fluidity, solidity, moistness and dryness] (ibidem, 329–348), form part of the first treatise of the *Abrégé*, *Des principes physiques* [*Of Physical Principles*], and of the second treatise, *Des Qualitez* [*Of Qualities*], respectively. In the second edition, published ten years later, these chapters are relocated (Bernier 1684, 2, Ch. X: 108–124), due to the addition of important sections in the overall structure of the *Abrégé*. The interest of the first chapter lies not only in its being a systematic development, largely taken over as such from the *Syntagma philosophicum*, but also in the fact that it engages with Descartes's arguments in the *Principles of Philosophy* in support of the idea that matter can be divided indefinitely. Bernier simplifies the corresponding chapter in the *Syntagma*[1] in the sense already indicated. But he also adds elements that seem original and that represent a direct response to Descartes. Thus he writes as a successor of Gassendi, taking over from him, and on his behalf, the banner of the fight of the atomists against the partisans of the continuum.

Bernier's line of argument starts from the observation that, in talking of matter, Descartes thinks in terms of a continuum, while in reality matter is of the order of the contiguous, that is, it presents itself as a juxtaposition of small bodies, better conceived of in terms of an aggregate [*amas*]. The division of matter, as Descartes sees it, would thus be accomplished in reality along the boundaries of the bodies that make up the aggregate and thus it is not a division of a continuum proper. Ultimately, the only continuous thing, properly speaking, would be the atom itself, which we must conceive as indivisible, at least as far as the laws of nature and God's ordinary power go, irrespective of miracles, which do not pertain to the science of physics. We cannot establish criteria that would distinguish among atoms (which are all solid, hard, continuous, indivisible) that would give one priority over the others, making one yield to the action of the other. By way of consequence, division would apply to the extended matter not in so far as it is continuous, but only in so far as it is discontinuous, that is, in so far as it is a juxtaposition of atoms.

In order to support this idea, Bernier reduces his investigation to the physical domain, and leaves geometry aside and thus bars the whole question of the infinitesimals, which had witnessed spectacular developments due to the work of Desargues and Pascal. The problem of the atoms thus becomes an exclusively physical problem and the arguments Bernier brings up in favor of the existence of atoms will consequently be uniquely physical arguments.[2]

Thus, for instance, Bernier thinks that the senses are too limited to allow perception of the contiguous in matter: they only afford us the sensation of a continuum. The presence and frequency of the void spaces in between the atoms of matter are placed in relation with the bodies' property of solidity, where solidity, or hardness, forms part of the very definition of the atom. The properties of physical bodies could be deduced from the behavior of the corpuscles that form the intimate structure of matter. In two of the chapters of the second treatise of the *Abrégé* that treat of the qualities, Bernier, following Gassendi (*Syntagma philosophicum*, Book VI, Chapters 7 and 8, in Gassendi 1658, 1: 402–409), provides a convincing articulation of the notions of dryness and of solidity, and of moistness and fluidity, respectively. Aristotle had done the same thing in *Of Generation and Corruption*, in reducing the qualities of the body to the sense of touch and establishing four primary qualities, that is, moist, dry, hot, and cold, from which all the other qualities are derived, either by subordination or by mixture. Aristotle claimed that the hard derived from the dry, and the soft from the moist, without elaborating on the difference between the solid and the liquid states (*Physics*, II, 2, 3–8, 330b 10).

Bernier starts from the observation that the dry is not as extended as the solid is, neither is the moist as extended as the fluid is, and thus establishes a species–genus relation between the two pairs of notions. After enunciating the Aristotelian definitions of the dry and the moist, he asks what it is that forms the fluid and what the solid; he thinks that there is, within the intimate structure of fluid bodies, a succession of corpuscles (which he sees, *contra* Descartes, as atoms) and of interstitial void spaces, alternating so as to permit the corpuscles to turn and roll around their own axes. The difference between the ways in which wheat and water may be said to "flow," in that whereas they both fill in the shape of the recipient in which they are poured—water seems continuous but wheat discontinuous—can thus be explained in terms of the difference between the sizes of the corpuscles and the surrounding void spaces characteristic of wheat and water, respectively.

The absence of interstitial spaces makes possible an explanation of the behavior of solid bodies. Due to their contiguous position and to their adherence, the corpuscles or atoms of solids cannot move for lack of room, which is one reason for the hardness of bodies in general. Coming back to the hypothesis that the fluid cannot be reduced to moistness, Bernier gives several examples of dry bodies that are nevertheless fluid, such as the air, molten metal, or quicksilver. They all flow, although they are not moist. This contrastive analysis of the ways in which moist and dry bodies flow finally permits a definition of moisture. A moist body is that which, in flowing, adheres with its parts to the body along which it flows, either by humidifying it on the surface or by impregnating it. The dry body, even when incorporated within fluid matter, is that which, in flowing, does not adhere with any of its parts, either at the surface or at the interior of the body along which it flows, but which, without loss or diminution of its substance, flows along it without humidifying it in any way.

Coming back to the first chapter, on the existence of atoms, let us note that there is a whole series of objections that Bernier brings to the physics of the *Principles of Philosophy*, although Bernier almost never mentions Descartes by name. These objections are not taken over from Gassendi who, in the *Syntagma*, does not deal in any substantial way with Cartesian physics. Bernier highlights the property of impenetrability, which Descartes did not maintain was an essential attribute of bodies, and associates it with those of hardness and solidity. On this ground, Bernier contests the model of the solid body furnished by the *Principles*, seen as a juxtaposition of particles in a state of rest relative to one another:

> I know very well that there are some who say that if several equal and very well polished cubes were perfectly arranged one on top of the other, they would form a mass at the interior of which there would be no interstitial void, but which could be easily divided, either by shaking or by hitting from one side. But they forget that the fundamental reason for the indivisibility of a mass is by no means so much the absence of small interstitial voids, as the solidity, hardness and unity of bodies, or their continuity; and that, conversely, the fundamental reason of divisibility is by no means so much the presence of small interstitial voids, as the plurality of bodies, or their aggregate, or discontinuity, if you will, or the mere contiguity of parts.
> (Bernier 1674, 1: 35)[3]

It is remarkable how Bernier reconstructs the Cartesian position of the *Principles* in order to better refute it with Gassendist arguments, while Gassendi himself never wrote on the subject. This discussion in itself is an argument for the status of Bernier as a true successor of Gassendi, in that he proves he is perfectly capable of identifying by means of philosophical analysis the positions that seem to him to be partially or completely vulnerable to the atomist attack.[4] This development, which is part of Bernier's piling of arguments in favor of the existence of atoms, is a distinctive sign of the battle that the Gassendist movement could no longer afford not to fight following the publication of the Cartesian physics. Gassendi himself had hesitated to engage this battle, as he was no doubt more engaged in mounting his attacks against the *Meditations* and against the *Responses to the Objections*. But Bernier takes it up twenty years later, and is able to combine an excellent understanding of atomist arguments with a reflection pointedly directed against Descartes's *Principles*:

> There is this difference between a continuous whole, or a whole whose parts are continuous, and an aggregate, whose parts are only contiguous, that the former is a simple, uniform body without interruption, and whose parts (if we call them that) are contained under one and the same surface or figure, while the latter is not so much a whole as an aggregate of several wholes, or of parts which, in so far as they are contiguous relative to one another, are not only distinct but also actually separated from one another and contained under several surfaces actually distinguished and separated, each of them having its own [surface] which is both particular and determined, besides the general [surface] of the whole.
> (Bernier 1674, 1: 35)[5]

The "cement" of the solid body is the very simplicity of the continuum, the idea of which could not be provided either by the movement or by the state of rest of the parts:

> Let no one say that there is no other difference between a continuous whole and an aggregate but that the parts of the aggregate are in movement while those of the continuum are in a state of rest; because it is neither motion nor rest that give us the idea of contiguity, no more than they do the ideas of softness, of hardness and of resistance of a body. And when we suppose that several little cubes like those we have described, simply arranged the one on top of the other, are in a state of rest, we will not fail to conceive that they are simply contiguous relative to one another, or that they do not form a continuous whole, in the same manner as we will conceive that they are solid, hard and resistant, even if they are in motion. Moreover, if there were no other distinction between the continuous and the contiguous, given that contiguity is subject neither to increase nor diminution, all bodies would be equally easy to separate, which is evidently contrary to experience: experience shows us that iron is harder to divide than wood, because iron is more solid or, which is the same thing, because its parts being more tightly packed with one another and having less interstitial small voids, it better approximates perfect solidity and continuity than wood does.
> (Ibid.: 36)[6]

Bernier uses here, against Descartes, Descartes's own method of the analysis of ideas, based on the presupposition of innate ideas and of the truth of clear and distinct ideas, which he had developed in the Fourth Meditation. Bernier borrows from Descartes the notion of the immutability of the laws of nature, and operates with a conception of nature which, while built on Cartesian foundations, already clearly points in the direction of the conception that will become the foundation of the mechanical philosophy:

> I will not say that one proof for the existence of atoms frequently invoked is that nature being always the same, that is, constant, firm and invariable in her works, she always produces the same species which she inviolably sustains and preserves in their natures, with their same force, growth and duration; and that she always impresses upon them the same differences and marks that distinguish them from other species; which she would certainly not do, did she not use principles that are certain and constant and by consequence indissoluble and immutable.
> (Ibid.: 36–37)[7]

Working with the opposition between the continuous and the contiguous, he represents matter as a system of continuous bodies grafted on the radical discontinuity of the alternation of atoms and voids. Differences in behavior during mechanical collisions are thus due to differences in the intimate organization of the matter of each body, that is, in the configuration of the atoms and the quantity of void in between the atoms.

If we now compare this with the second revised edition of the *Abrégé*, of 1684, we will notice that this long elaboration, which takes Descartes as a target without naming him, all but disappears. There are nevertheless two references to Descartes in this edition. First of all, there is an argument in favor of the infinitely small taken over from the *Logique de Port-Royal*, where Arnauld and Nicole, despite their Cartesianism, seem to be indebted to Pascal:

> What means do we have, says the author of the Logique, usually called the Logique de Port-Royal, to understand *that we can never come to a part so small that it does not enclose not only several other parts, but an infinity of them; that the smallest grain of wheat contains in itself as many parts, even if proportionally smaller, as the whole world does; that all the imaginable figures are actually included in it and that it contains in itself a little world with all its parts, a sun, a sky, stars, planets, an earth, in admirably just proportions; and that there is no part of this grain that does not itself contain another, and then another proportional world*. This is how this author reasons; but the surprising thing is that, instead of rejecting an opinion which he admits entails extremely incomprehensible and strange, not to say extravagant, things, he approves it and believes in this infinity of actual and real worlds contained in the millionth, or if you like in the milli-millionth part of the extremity of the foot of a mite.
>
> (Bernier 1684, 2: 117–118)[8]

A second attack is borrowed from Gassendi and is aimed at Descartes's indefinite divisibility of matter; but Descartes is not named here (and only features as "one of our moderns"[9]), while the first edition did mention his name in full (Bernier 1674: 42).

Thus, the text of the *Abrégé* goes through remarkable modifications from one edition to the other. On the one hand, the 1684 edition is much more obviously indebted to Gassendi's *Syntagma*. On the other, Bernier adds his *Doutes*, where he distances himself to some extent from Gassendi's doctrine. His justification, addressed to Mme de la Sablière, reads: "It's been thirty to forty years since I've been utterly convinced of certain things in my philosophy, but now I'm starting to have doubts." These doubts show that Bernier is highly aware of the debates of the time around the questions of space, place, and movement. Adopting a nominalist position, he comes quite close to Pascal's view that it is useless to define "primitive words"[10] that are to be seen as "modes" of substance (*Doute IV: Si le mouvement se peut, ou se doit définir?*, in Bernier 1684: 405–413) that are already employed to clarify the substance itself and which could not thus be further defined. The tone of these doubts, which he dedicated to Mme de la Sablière, is one of tempered skepticism. Gassendi himself is taken to task for having introduced the notion of space as an "eternal, incorporeal, immense, independent and immovable being" (ibid.: 410), a notion that stands as a threat to the idea of God. "I ask, can a philosopher admit with ease a being other than God that is incorporeal, immense, independent and incorruptible, or incapable of being destroyed?" (*Doute I: Si l'espace de la manière que Monsieur Gassendi l'explique est soutenable*, ibid.: 382). The only reason Gassendi had introduced this notion of space seems to have been to ground the notion of the immovability of place, "in favor of which he was prejudiced, together with most of the ancients" (ibid.). In support of this view, Bernier claims that the void can be conceived, without recourse to the notion of a space that would fill in its dimensions, as pure privation (ibid.: 385), which requires neither substance nor accident. If he quotes Emmanuel Maignan on the possibility of reconciling the existence of an incorporeal space with the Aristotelian place,[11] he immediately refutes this idea, on the grounds of the principle of the nonmultiplication of entities.

Bernier defines place as "the surface of a surrounding body" (*Doute II: Si l'on peut dire que le lieu soit l'espace?*, ibid.: 394), whereby he wants to express the situation of a body "located in such a place" by means of the extrinsic denomination of that surface. He raises against himself Aristotle's objection that in this way the outermost sphere would hold no place. But there's nothing to it! This objection only gains its force from a "preoccupation" of the mind, since all the things we can see are bounded by a surrounding body. We have to admit that "there is no body, no thing beyond the universe" (ibid.: 397). Lucretius's supposition that an arrow that would cross over the border of the finite universe would be in no place, should be resolved thus: the arrow would go beyond the universe, but it would be in no place, and by consequence it would not be in the void either, since for there to be a void, there must be place as well (ibid.: 398). The incorporeal and penetrable extension, space, and internal place, are therefore nothing more than chimeras, imaginary notions without any reality.

In discussing the supposed immovability of place, Bernier invokes in passing a definition of motion that can be considered Cartesian: "a successive application of a body to the parts of the neighboring bodies." Once motion is thus defined, it follows that an immovable thing, such as a tower, in the middle of fluid air, would move, and so would a fish holding up against a flow of water, but that a moving body, such as a fish carried by the water flow at the interior of a block of ice, would not move (ibid.: 401). If instead we do not accept this definition and simply consider the relation with a place, we will hold that the tower changes places but does not move, and that the fish carried away within a block of ice moves but does not change places. Gassendi's solution, which is to define motion as "the passage from place to place, that is, a successive application of the moving body to the various parts of the space," is not appropriate either, since it presupposes an immovable space: "But the remedy seems even worse than the disease, since in order to confer immovability to a place and to be able to define motion, he is forced to have recourse to space, that is to say, to a thing which does not exist, or, as we have shown, which is no more than pure fiction, or a purely imaginary being" (Ibid.: 402–403).[12]

Having examined Bernier's *Doutes*, we can now confirm the results of the analysis of the two chapters undertaken above. Bernier selected different passages from Gassendi's work for inclusion in the two editions of the *Abrégé*. He also reworked them and added his own comments to them, and left out other passages. He reshaped some of the arguments to answer his time's concerns, and capably used material from the partisans of the void, be they overt, like Maignan, or dissimulated, like Pascal. He contradicted Gassendi on several fundamental presuppositions, such as the existence of an immovable and penetrable space, in order to better defend the atomist theses against the triumphant Cartesianism of the time.

As historians of science have repeatedly pointed out, a critical edition of the *Syntagma* is long overdue. But we can also deplore the fact that the modern reprint of Bernier's *Abrégé* has only followed the 1684 edition without signaling the remarkable modifications—the additions, the deletions—relative to the first edition, and without prompting discussion around the immense task Bernier took up in the service of European Gassendism. Even if it sometimes conveyed smuggled goods, we should not forget that the *Abrégé* was a warship. And like all means of military intervention, it mounted up, under the cover of anonymity, an armory of important choices and intelligence operations.

Translation by Sorana Corneanu

NOTES

1. *De opinione statuentium materiam solas atomos insectiliave corpuscula praedita magnitudine, figura, pondere dumtaxat*, in *Physica*, Section I, Book III, Chapter V, in Gassendi 1658, 1: 256–266.
2. Walter Charleton had proposed a similar approach in Charleton 1654. In Book II, Ch. 2—*An liceat in materiam physicam, sive sensibilem, transferre Geometricas Demonstrationes?*—he argued that the methods employed by the mathematicians in tackling the continuum are completely useless for an examination of the physical body.
3. Bernier 1674, 1: 35: "Je sais bien encore qu'il y en a qui disent que si plusieurs petits cubes égaux et très polis étaient bien arrangés les uns sur les autres, ils seraient une masse au dedans de laquelle il n'y aurait aucun vide intercepté et qui cependant se pourrait aisément diviser, soit en la secouant, soit en lui donnant quelque petit coup par le côté. Mais ils ne prennent

pas garde que la raison primitive de l'indivisibilité d'une masse n'est point tant l'absence des petits vides interceptés, que la solidité, la dureté et l'unité de corps, ou la continuité; et qu'au contraire la raison primitive de la divisibilité n'est point tant la présence des petits vides interceptés, que la pluralité de corps, ou l'amas, la discontinuité, s'il est permis de parler de la sorte, ou la simple contiguïté de parties."

4. See also another original contribution by Bernier in response to the Cartesian cosmology of the *Principles* in Ch. IX, *De la nécessité des petits vides entre les corps*, pp. 89–92 in the 1674 edition of the *Abrégé*, enlarged in the 1684 edition at pp. 184–190.

5. Bernier 1674, 1: 35: "Or il y a cette différence entre un tout continu, ou dont les parties sont continues, et un amas, dont les parties ne sont que contiguës, que le premier est un corps simple, uniforme, sans interruption, et que toutes ses parties (si on peut les appeler parties) sont contenues sous une seule et unique superficie ou figure; au lieu que le second n'est pas tant un tout qu'un amas de plusieurs touts, ou de parties qui n'étant que contiguës entre elles, sont non seulement distinctes, mais séparées actuellement les unes des autres et contenues sous plusieurs superficies actuellement distinguées et séparées, chacune ayant la sienne propre et particulière et déterminée, outre la générale du tout."

6. Ibid.: 36: "Et qu'on ne dise point qu'il n'y a que cela de différence entre un tout continu et un amas, que les parties de l'amas sont en mouvement et celles du continu en repos; car ce n'est ni le mouvement ni le repos qui nous donnent l'idée de la contiguïté, non plus que de la mollesse, ou dureté, et résistance d'un corps; et quand on supposerait que plusieurs petits cubes tels que nous avons dit, et arrangés simplement les uns sur les autres, fussent en repos, nous ne laisserions pas de concevoir qu'ils ne seraient que contigus entre eux, ou qu'ils ne seraient pas un tout continu comme nous concevrions qu'ils seraient solides, durs, et résistants en soi, quoi qu'ils fussent en mouvement. Outre que s'il n'y avait point d'autre distinction entre la continuité et la contiguïté, comme la contiguïté ne reçoit point de plus, et de moins, tous les corps seraient également aisés à être séparés, ce qui est évidemment contraire à l'expérience qui nous montre que le fer se divise plus difficilement que le bois, parce que le fer est plus solide, ou ce qui est le même, parce que ses parties étant plus serrées et ayant moins de petits vides interceptés, il approche plus de la solidité et continuité parfaite que le bois."

7. Ibid.: 36–37: "Je ne vous dirai point qu'on apporte ordinairement pour preuve de l'existence des atomes que la nature étant toujours la même, c'est-à-dire constante, ferme et invariable dans ses ouvrages, elle produit toujours les mêmes espèces qu'elle entretient et conserve inviolablement dans leur nature, dans leur même force, accroissement et durée; et qu'elle leur imprime toujours les mêmes différences et marques qui les distingue des autres espèces; ce qu'elle ne serait assurément pas, si elle ne se servait de principes certains et constants, et par conséquent indissolubles et immuables."

8. Bernier 1684, 2: 117–118: "Quel moyen de comprendre, dit l'auteur de la Logique ordinairement appellée la Logique de Port-Royal, *que l'on ne puisse jamais arriver à une partie si petite, que non seulement elle n'en enferme plusieurs autres, mais qu'elle n'en enferme une infinité; que le plus petit grain de blé enferme en soi autant de parties, quoiqu'à proportion plus petites, que le monde entier; que toutes les figures imaginables s'y trouvent actuellement et qu'il contienne en soi un petit monde avec toutes ses parties, un Soleil, un Ciel, des Etoiles, des Planètes, une Terre*

dans une justesse admirable de proportions; et qu'il n'y ait aucune des parties de ce grain qui ne contienne encore de même un autre monde, et ainsi à l'infini, sans qu'il s'en puisse jamais trouver aucun dans laquelle il ne se trouve toujours un nouveau et puis un nouveau monde proportionnel. C'est ainsi que raisonne cet auteur; mais ce qu'il est surprenant, c'est qu'au lieu de rejeter une opinion d'où il avoue qu'il suit des choses si fort incompréhensibles et si étranges, pour ne dire pas extravagantes, il l'approuve et croit cette infinité de mondes actuels et effectifs dans la millième ou si vous voulez dans la mille-millième partie de l'extrémité du pied d'un ciron." This paragraph did not exist in the first edition.

9. Ibid.: 120–121. He follows Gassendi, *Physica*, section I, liber III, Ch. V of the *Syntagma*, in Gassendi 1658: 263.
10. Blaise Pascal, *De l'esprit géométrique* (c. 1657), in Pascal 1991: 395, some of the propositions of which are repeated by Arnauld and Nicole in the *Logique de Port-Royal*, 1662.
11. He seems to think of Maignan 1653, in particular of Ch. 8: *De corpore physico locato*, proposition 3: *Ad immobilitatem, quae de conceptu loci est, explicandam sufficit immbilitas spatii imaginarii apprehensa*, Bernier 1684: 734–735.
12. Ibid.: 402–403. The Cartesian definition of motion: "une application successive d'un corps aux parties des corps voisins;" Gassendi's definition of motion: "le passage de lieu en lieu, c'est-à-dire une application successive du corps mobile aux diverses parties de l'espace;" Bernier's comment: "Mais le remède me semble être pire que le mal ; puisque pour donner l'immobilité au lieu et pouvoir définir le mouvement, il est obligé d'avoir recours à l'espace, c'est-à-dire à une chose qui n'est point, qui n'existe point, ou, comme nous avons montré, qui n'est qu'une pure fiction, ou un être purement imaginaire."

BIBLIOGRAPHY

Ariew, R. (1992) "Bernier et les doctrines gassendistes et cartésiennes de l'espace: réponses au problème d'explication de l'eucharistie," *Corpus*, 20/21: 155–170.

Aristotle. (1984) *The Complete Works of Aristotle*, 2 vols, ed. J. Barnes, Princeton: Princeton University Press.

Arnauld, A. and Nicole, P. (1996) *Logic or the Art of Thinking*, 4th edn [1683], ed. J. V. Buroker, Cambridge: Cambridge University Press. First published in French in 1662.

Bernier, F. (1674) *Abrégé de la philosophie de M. Gassendi*, 1st edn, Paris: Jacques Langlois et Emmanuel Langlois.

——. (1684) *Abrégé de la philosophie de Gassendi*, seconde édition revue et augmentée par l'auteur, Lyon: Anisson, Posuel & Rigaud.

——. (1992) *Abrégé de la philosophie de Mr Gassendi*, Paris: Fayard, 7 vols.

Charleton, W. (1654) *Physiologia Epicuro-Gassendo-Charltoniana*, London.

Corpus. (1992) *Corpus. Revue de philosophie*, 20/21, *Bernier et les gassendistes*.

Descartes, R. (1996) *Œuvres de Descartes*, revised edn, 11 vols., eds. C. Adam and P. Tannery, Paris: Vrin.

Gassendi, P. (1658) *Opera omnia*, 6 vols, Lyon: Laurent Anisson et Jean-Baptiste Devenet.

——. (1955) *Pierre Gassendi—sa vie, son oeuvre: 1592–1655*, ed. S. Murr, Centre international de synthèse, Paris: Albin Michel.

Koyré, A. (1980) "Gassendi et la science de son temps" in *Etudes d'histoire de la pensée scientifique*, Paris: Gallimard.

Maignan, E. (1653) *Cursus philosophicus*, 1st edn, vol. 2, Toulouse.

Murr, S. (1992) "Bernier et le gassendisme," *Corpus*, 20/21: 115–135.

———. (1997a) *Gassendi et l'Europe (1592–1792)*, Actes du colloque international de Paris, *Gassendi et sa postérité*, Sorbonne, 1992, Paris: Vrin.

———. (1997b) "Bernier et Gassendi: Filiation déviationiste?" in ed. S. Murr 1997a, 71–114.

Pascal, B. (1991) *Oeuvres complètes*, texte établi, présenté et annoté par Jean Mesnard, Paris, Desclée de Brouwer, tome III.

4 The Matter of Medicine
New Medical Matter Theories in Mid-Seventeenth-Century England

Peter R. Anstey

The mechanical philosophy of the seventeenth century is not normally associated with medicine. To be sure, some mechanical philosophers developed mechanistic accounts of animal physiology, most famously Descartes in *De homine* (1662, see Descartes 1998) and later Borelli's *De motu animalium* (1680–1681). And certain specific physiological structures came in for close analysis by mechanical philosophers and physicians alike. Here again one thinks in the first instance of the thirty-year debate over the Cartesian theory of the motion of the heart and its final resolution in 1669 in the work of the physiologist John Mayow.[1] But mechanistic physiology did not have much discernible influence on curative medicine, that is, on the theory and practice of physic.

There is one facet of the mechanical philosophy, however, that did have an important impact on medicine, an impact that has hitherto been underappreciated, and that is matter theory. As it happens, from the early 1650s and then throughout the 1660s and beyond, the new corpuscular matter theory promoted by the likes of Robert Boyle began to make inroads into English medicine. In fact, the new corpuscular matter theory, I will argue, had some very important and uncomfortable implications for the theory and practice of physic in mid-century England. Debate over matter theory was an important strand in the heated medical debates of this period, a period that is one of the most fraught in the history of Western medicine.

This chapter aims to illustrate the different ways in which physicians and medically oriented natural philosophers responded to the new corpuscular matter theory. But before we turn to the matter of medicine there is some scene setting to do. For debate over matter theory among physicians did not arise spontaneously, but rather emerged as the result of important earlier developments. Some of these were new trends and advances within medicine itself and others were derived from the new natural philosophy.

BACKGROUND

First, it is well known that from the time of Paracelsus (d. 1541) and his followers the call for the reform of medicine could not be ignored in the main centres of Europe. The message was amplified by Francis Bacon in England and by the leading chymists of the late sixteenth century and early seventeenth century such as Joan Baptiste van Helmont (1579–1644) and Daniel Sennert (1572–1637). By the 1650s in England, the pressure for an instauration in physic was increasing in intensity among the group of competent, outspoken, and politically astute chymical physicians. All of this has been well documented over the last fifty years and is now relatively well understood.[2] Just what needed to be reformed in traditional medicine will become apparent below.

Second, as is well known, there had been some landmark developments in the understanding of animal physiology in the first half of the century that had implications for traditional medical theory and, to a lesser extent, for medical practice. Harvey's discovery of the circulation of the blood, the discovery of the thoracic duct, the discovery of the lymphatic system, and so on were both exciting and unsettling for those practicing traditional medicine in the middle of the seventeenth century. They were exciting, in part, because they presaged more discoveries of their kind, and indeed discoveries proceeded apace in the very period on which we are focused here. And they were unsettling, in so far as they served to undermine traditional medical teaching about anatomical structures and physiological systems on which Galenic method was based, and they heightened the sense among the learned establishment that there is much that is not known and much that is relied upon that might well prove to be unfounded.

Third, and perhaps most important for our concerns here, is the fact that a new emphasis on observation and experiment was emerging among the medical fraternity, an emphasis that was paralleled among natural philosophers and that is almost certainly the salient methodological shift in early modern medicine, a shift that was to have ramifications for medical methodology well beyond our period of interest. One seam of this new emphasis on observation and experiment is the Harvean experimental approach that was pursued by the Oxford physiologists.[3] Indeed, Harvey's emphasis on observation and experiment as spelled out in the Preface to *Exercitationes de generatione animalium* of 1651 is very revealing for anyone interested in the new medical methodologies of the period. Another seam is, of course, the experimental work of the chymical physicians. They were the most ardent supporters of observation over tradition and speculative theory in the medical debates of the period.

So we have the broad-based call for medical reform, we have the challenging and in some respects unsettling discoveries in physiology, and we have a new emphasis on observation and experiment. These developments were some of the factors that, in a sense, loosened up the soil of physic in which the matter theory of the new philosophy could take root. Let us turn then to the matter of medicine in Britain in the 1660s. The place to start is with a rough sketch of the traditional hegemonic theory, Galenism, which came under threat from the developments outlined above.

GALENISM AND THE FIRST FOUR QUALITIES

As we all know, Aristotelian matter theory posited four elements: earth, air, fire, and water. Now, according to this four-element theory, all material bodies are made up of one or more of the elements. Material bodies can change in so far as the relative quantities of these elements change within them. But the elements can change too. They can transmute into each other. Both sorts of change, that is, changes in material substances and changes in the elements themselves, are explained using the Aristotelian theory of qualities.

There are four primary qualities that form two pairs of opposites: hot/cold, wet/dry. By varying the predominance of one or more of these qualities the mix of elements in the body changes. Thus, by adding heat to water we get air, and so on. Furthermore, in the Aristotelian theory of qualities there are also secondary qualities that are also two pairs of opposites, heavy/light and dense/rare, and a whole host of other divisions among the qualities of bodies. All of these divisions in the hierarchy of qualities are explained, ultimately, in terms of the *primae qualitates*. So the Aristotelian primary qualities are primary in at least two senses: they are the primary *explanans* of the characteristics of bodies and they are the ontological ground of these characteristics. If a quality is inexplicable in terms of the *primae qualitates* it is classed as an occult quality. There is nothing mysterious or sinister about occult qualities, they are just inexplicable by the usual *explanans* of the theory, that is, hot, cold, wet, and dry.

THE GALENIC THEORY OF DISEASE

Now the Aristotelian theory of qualities was used to explain the constitution of the four bodily humours in the Galenic humoral theory of disease. The four humours, blood, black bile, yellow bile, and phlegm, all differ relative to the predominance of the four primary qualities. Each humour combined two primary qualities: yellow bile was hot and dry; black bile was dry and cold; blood was wet and hot; and phlegm was wet and cold. Disease was understood to result from an imbalance or disequilibrium of the natural ratios of these bodily humours within an individual body. Every different animal body had its own unique natural equilibrium or temperament of the humours, and so every different body required its own unique regimen in order to restore its natural temperament.

Physic as practiced by Galenic physicians was divided into physiological, pathological, semiotical (diagnostic), hygenical, and therapeutical parts.[4] The primary symptoms of any disease, while traceable to the imbalance in bodily humours, were determined in terms of the primary qualities. The determination of the presence or absence of the primary qualities was the main diagnostic tool of the physician. Furthermore, the cure of disease was understood in terms of adjusting the humoral balance by changing the relative weightings of the primary qualities. It is hardly surprising then, when attacking the Galenists in his *Compleat Body of Chymistry* (1664), that Nicaise Le Febvre can speak of

> those which are only Physitians by name, and who after they have perused some University Writings, which do perswade them, that, Physick is nothing else but an art of discerning heat and cold, immediately take upon themselves the practice, and fill their discourses with nothing else but notions of Heat and Cold; and all their skill tends to speak upon the more or less of these qualities.
>
> (Le Febvre 1664: 108)

It is clear then that the Aristotelian theory of matter and qualities was absolutely fundamental to the Galenic system of medicine. Well-being, illness, and cure were all explained in terms of the Aristotelian theory of qualities. Just how deeply connected the theory of qualities was with the practice of physic can be seen in the doctrine contraries.

It was widely believed that a key to the cure of disease was to use medicines and therapies that induced the opposite quality to that which presented itself. The motto of this doctrine was *Contrariorum contraria sunt remedia*. It seems commonsensical that if the patient is hot then they should be cooled, if cold, then the physician should warm them up. But the Galenic physicians did not stop there and the interpretation of the doctrine of contraries led to many bizarre and dangerous treatments. As Boyle tells us in his *Usefulness of Natural Philosophy, II section i* (hereafter *Usefulness of Natural Philosophy*),

> diverse famous, and, otherwise learned Doctors, . . . have believ'd and taught that the Stone of the Kidneyes is produc'd there by slime baked by the heat and drinesse of the Part; as a portion of soft Clay may, by externall heat, be turn'd into a Brick or Tile. And accordingly they have for cure, thought it sufficient to make use of store of Remedies to moisten and cool the Kidneys.
>
> (Boyle 1999–2000, 3: 313)

It also had implications for the types of herbal and chymical remedies that were prescribed.

OPPOSITION TO GALENISM

The writings of Joan Baptista van Helmont were the most important source for opposition to Galenism in England, and the Helmontian chymists of the 1650s were vehement in their opposition to the Galenic "Methodists." Marchamont Nedham's presentation of van Helmont's succinct summary of Galenic physic captures the spirit of this opposition.

This *Galen* arrogating to himself the glory of such as went before him, inlarged the Art of Physick, which before was contained in a few Rules, unto vast Volumes. He was pleased forsooth to determine, according to *Hippocrates*, that all Bodies were made up of *four Elements*; and parallel to these he established *four Qualities*, and so many *simple Complexions*, and then *as many Couples of complex Qualities*, and thence he confirmed also *four Humors* to make up our *Constitutions*; which were devices *that others had dreamt of before him*: And then, out of the strife and disagreement of these, as well simple, as complicate with fictitious humors, he would needs derive almost all Diseases, with the Scopes and Indications of Cure; as on the other side he concluded Health to arise from their Agreement and Proportion with each other. Moreover, every Disease he declared to be but a mere Disposition in quality; and that *Contraries were to be cured by Contraries*.
(Nedham 1665: 240)

The doctrine of contraries was just one facet of Galenism that came under attack from the chymical physicians. We even find the young John Locke prefacing his unpublished disputation on the uses of respiration (c. 1666) with the claim:

Whether contraries are cured by contraries? I deny it.
This thing, namely that contraries are cured by contraries, has been greatly vaunted by authors as if it were the foundation of the whole practice of medicine and established by the consensus of the ancients and by experience.
(National Archives PRO 30/24/47/2, fol. 71v)[5]

And we find fierce attacks on the doctrine in the writings of George Thomson such as his chapter in *Galeno-Pale* of 1665 entitled "Of that fictitious Rule of Contraries, by which the Dogmatists are guided in the cure of Diseases."[6]

But our interest here is on something even more fundamental, something of which this opposition to the doctrine of contraries is merely a symptom, and that is the emergence of a fully fledged theory of qualities that rivalled and eventually displaced the Aristotelian theory that underlay the Galenic *methodus medendi*. By the mid-seventeenth century, the Aristotelian theory of qualities was under serious challenge from the theory of qualities that had its origins in Greek atomism and that had persisted in some corpuscular matter theories and received significant polemical deployment in the writings of Descartes, Gassendi, and others in the early seventeenth century. This was the view that the shape, size, and motion of corpuscles or atoms were the main explanatory categories in the theory of qualities and that rather than being the *explanans*, the traditional *primae qualitates* were to be explained in terms of shape, size, and motion.

It is perhaps no coincidence that physicians were some of the first conduits for the transmission of Gassendist and Cartesian ideas into England in the late 1640s and early 1650s. One might instance the early publications of the physician Walter Charleton, whose *Physiologia Epicuro-Gassendo-Charltoniana* (1654) provided the first English paraphrase of Gassendi's atomism.[7] Likewise the Dorset physician Nathaniel Highmore's "The Sympathetical Cure of Wounds," which was appended to his *The History of Generation* (1651), and which was dedicated to Boyle, gives a summary of an atomistic or corpuscular matter theory and applies it to the explanation of the efficacy of the weapon salve.[8]

Now it is tempting to regard the situation as involving a new set of *primae qualitates* that had more explanatory power than the traditional four. But this is not quite correct as a characterization of the early stages of the penetration of the corpuscular theory. It is true, however, of Locke's development of the primary and secondary quality distinction in his *Essay concerning Human Understanding* (1690). Locke's distinction should be regarded as the decisive step in consolidating the primacy of the new primary qualities and articulating their relation to the sensible qualities. In fact, Locke's apparent freedom to call shape, size, and motion primary qualities is, perhaps, indicative that the situation had changed so radically in matter theory by 1690 that the old theory of qualities no longer had proprietorial rights over the term "primary quality" (Locke 1975: II. viii). Importantly, however, his predecessor Robert Boyle never used the term in this way and refrains from calling shape, size, and motion qualities at all.

Our interest here is in the manner in which physicians coped with and reacted to the new corpuscular principles. But first we need to see the route by which the corpuscular principles were introduced into debate about the principles of physic more generally and the singularly most important point of penetration is in the writings of Robert Boyle.[9]

ROBERT BOYLE

Boyle had long maintained an interest in physic. His first publication was entitled "An Invitation to a free and generous Communication of Secrets and Receits in Physick" (1655, Boyle 1999–2000, 1: 1–12). Much has been written on his early interest in medicine and its co-evolution with his work as a natural philosopher. Not only was he closely associated with one of the more outspoken critics of Galenic medicine in the 1650s, the American George Starkey, but his involvement in the physiological researches in Oxford, his connections with the chymical physicians more generally, his own practice of chymistry, his broader natural philosophical interests, and personal connections with the medical fraternity meant that he developed a keen interest in the current state of the theory and practice of physic.[10] His early *Sceptical Chymist* contained some material of relevance to medicine, but it was his *Usefulness of Natural Philosophy* of 1663 that marked his entry into serious engagement with the medical fraternity and which, perhaps inadvertently, drew him into the fray. This work will be our point of entry into Boyle's published views on the Galenic principles and his corpuscularian alternative. Boyle's unpublished views in his suppressed critique of Galenism are more revealing, but were not part of the medical debate of the day (see Hunter 2000c).

Boyle calls the four elements, four humours, and four first qualities the Galenic *principles*. He is at a loss to see how the Galenists can explain the operation of medicines using their principles, but he has great hopes for the efficacy of the principles of the corpuscular philosophy. This is nicely captured in his discussion of the operation of specific medicines in his *Usefulness of Natural Philosophy*:

> For I see not how from those narrow and barren Principles of the four Elements, the four Humours, the four first Qualities (and the like;) Effects, far lesse abstruse then the Operations of Purging Medicines, can satisfactorily be deduc'd. Nor can I find, that any thing makes those Physitians, that are unacquainted with the Philosophy that explains things by the Motions, Sizes, and Figures of little Bodies, imagine they understand the account upon which some Medicines are Purgative, others Emetick, &c. And some Purgative in some Bodies, Vomitive in other, and both Purgative and Vomitive in most; but because they never attentively enquire into it.
>
> But (which is the next thing I have to represent) if we duely make use of those fertile and comprehensive Principles of Philosophy, the Motions, Shapes, Magnitudes and Textures of the Minute parts of Matter, it will not perhaps be more difficult to shew, at least in general, that Specificks may have such Operations, as are by the judicious and experienc'd ascrib'd to them, then it will be for those that acquiesce in the vulgar Principles of Philosophy and Physick, to render the true Reasons of the most obvious and familiar operations of Medicines.
>   (Boyle 1999–2000, 3: 465–466)

He also criticises those physicians who, realising the barrenness of the Galenic principles, substitute principles of their own. On the whole these new principles are the *tria prima* of Paracelsus or some variant. He replaces them with his own principles of the corpuscular philosophy as he tells us in *Forms and Qualities:*

> And though the Unsatisfactoriness and Barrennesse of the School Philosophy have perswaded a great many Learned Men, especially Physicians, to substitute the Chymists Three principles, instead of those of the Schools; and though I have a very good opinion of Chymistry it self, as 'tis

a Practicall Art; yet as 'tis by Chymists pretended to contain a Systeme of Theoricall Principles of Philosophy, I fear it will afford but very little satisfaction to a severe enquirer, into the Nature of Qualities. For besides that, as we shall more particularly see anon, there are Many Qualities, which cannot with any probability be deduc'd from Any of the three Principles; those that are ascrib'd to One, or other of them, cannot Intelligibly be explicated, without recourse to the more Comprehensive Principles of the Corpuscularian Philosophy. To tell us, for instance, that all Solidity proceeds from Salt, onely informing us, (where it can plausibly be pretended) *in what* materiall principle or *ingredient* that Quality *resides*, not *how* it is *produced*; for this doth not teach us, (for example) *how* Water even in exactly clos'd vessels comes to be frozen into Ice; that is, turn'd from a fluid to a Solid Body, without the accession of a saline ingredient.

(Boyle 1999–2000, 5: 301)

These corpuscularian principles are the shape, size, motion, and texture of the constituents of matter. The term "principle" here denotes more than an explanatory category, it denotes a metaphysical category as well. This is why Boyle refuses to call shape, size, and motion qualities, for they are the replacement principles of the new natural philosophy. They are not qualities but rather they are, following Descartes, the modes of matter and as such they are the ontological ground upon which the theory of qualities is to be based. Hence in the period in which the corpuscular philosophy was establishing itself, it is wrong to say, at least if one follows Boyle, that the corpuscular philosophy offered a new theory of qualities to replace that of the Schools and the Galenists if one thinks of shape, size, and motion as the new explanatory qualities. For the new theory of qualities is predicated on these mechanical principles or modes.

Boyle's *Usefulness of Natural Philosophy* was full of recommendations and experiments that were of relevance to physicians and chymists alike. It was not directed at any school in particular and the work is characterized by Boyle's usual diffidence and *via media* approach. However, it was soon to become a central text in the heated debate over the status of Galenism, not least because of the use made of the text by the controversialist and physician Marchamont Nedham.

MARCHAMONT NEDHAM (1620–1678)

Arguably the most influential attack on Galenic medicine in the 1660s was Marchamont Nedham's *Medela medicinae* of 1665, which announces itself on the title page as

> A Plea for the Free Profession, and a Renovation of the Art of Physick ... Tending to the Rescue of Mankind from the Tyranny of Diseases;

and of Physicians themselves, from the *Pedantism* of old *Authors* and present *Dictators*.

Nedham's *Medela* attacks Galenism with a vengeance and not surprisingly stirred up the ire of the Galenic physicians.[11] Replies soon came from Robert Sprackling, Henry Stubbe, John Twysden, and George Castle among others. Now Nedham's polemical strategy, a common ploy in medical texts from this period more generally, is to string together quotes from other medical and natural philosophical authorities. Not surprisingly two of the prominent authorities he calls upon are Thomas Willis and Robert Boyle.

In the seventh chapter he attempts systematically to dismantle the four central claims of Galenism. In section one he denies the doctrine of the four elements: "[a]way with the frigid Notion of *four Elements*, which he, out of *Aristotle*, makes to be Principles of all mixt Bodies" (Nedham 1665: 243). He begins by citing the leading physician Thomas Willis, who claims that these principles (and the first four qualities) are only adequate to explain things "in gross." And he follows Willis in substituting for the traditional four elements the five elements of the chymists, spirit, sulphur, salt, water, and earth (ibid.). These principles are not simple or uncompounded, but are merely those principles into which all natural things are resolved. Willis, and Nedham following him, leave room for a homogenous matter underlying the chymical principles. Moreover, these principles are the ones that bodies are resolved into by the application of fire in chymical experiments. And here Nedham appeals to Francis Bacon's authority (ibid.: 244).

Second Nedham dismisses the first four qualities: "After the *four Elements*, away also with the *four Qualities* which are attributed thereto" (ibid.: 245). Following another passing allusion to Bacon, Nedham appeals to van Helmont, who has shown "what mischief the admitting of these causeth in the cure of Diseases" (ibid.). Van Helmont's claim against wet and dry is that they are not strictly qualities in that they do not admit of degrees in so far as water never dries and dryness is really a privation of moistness. Moreover, as that ancient authority Hippocrates says, diseases can be caused by acid, bitterness, and saltiness in addition to the first four qualities (ibid.: 246–247). Nedham then sums up his position as follows:

> the Frame of Galenick Physick seems (for the most part) to be built upon *Hot* and *Cold*, and a attemperation [regulating] of these with *Moist* and *Drie*, therefore both the Method and Medicins devised in proportion to those Conceits, must needs be erroneous and insufficient,

and has a jab at the doctrine of contraries for good measure:

> and jumbling of pretended Contraries, to bring them to agreement and concord, as it were by Mathematical Rule of proportion: That is to say, the whole Fabrick of Natural Science hath hitherto stood upon mere

> opinions obtruded on us by gross Heathens, upon whom it is a shame to hear what high Elogies are bestowed.
>
> (Nedham 1665: 248–249)

Interestingly, he cannot resist an extended quotation from Boyle, though he omits to mention that Boyle not only dismissed the Galenic principles, but also the chymical ones as well (ibid.: 250–251).[12]

Thirdly, having dismissed the Galenic doctrine of the four qualities, Nedham feels he can summarily dismiss the four temperaments as well in virtue of the fact that the latter is predicated upon the former (ibid.: 257). So he turns finally to the four humours which "Conceit" he goes on to dismiss using similar argument forms.

Now in addition to this comprehensive dismissal of the principles of Galenic medicine, Nedham had his own theory of disease. It is the somewhat bizarre theory that many diseases are caused by wormatic matter or worms, which are akin to seminal principles and which are found in rotting flesh and in putrid water. Nedham aligns himself with the new experimental philosophy and furnishes his readers with experimental evidence for his own theory of disease (ibid.: 191ff and 255). All of this is indicative of the newfound freedom that reformist physicians felt when approaching the question of the nature of disease. There were many seminal theories of disease developed at this time such as those of George Thomson and John Locke.[13] But what is of interest here is Nedham's advocacy of the theory that many fevers and other diseases are caused by airborne morbific particles. This miasmic view was promoted by Boyle and it is interesting to note that on a number of occasions in describing these morbific particles Nedham uses the terms "atom" and "corpuscle" (Nedham, 1665: 116, 120, 122, 180, 195). Thus while his attack on Galenic theory appeals to the five chymical principles, and he certainly does not use the explanatory efficacy and economy of the corpuscular principles against the Galenists, his own view, or better, the view he derives from Willis, is broadly corpuscularian.

How did the Galenists respond to this critique of the four elements, four qualities and four humours? An obvious place to start is with the Oxford-trained physician George Castle.

GEORGE CASTLE

George Castle (1634/5–1673) went up to Oxford in the same year as John Locke, 1652, and his intellectual trajectory was somewhat similar to that of Locke in the mid-1660s. He took an MD in 1665 and became a fellow of the Royal Society in February 1669. He was never a Fellow of the College of Physicians. His *The Chymical Galenist* of 1667 is, as its title suggests, a defense of Galenism in the light of the new philosophy. But it is also a pointed attack on Nedham's *Medela medicinae*. In chapter six Castle defends the principles of Galenic medicine as being based upon experience and not speculative hypothetical systems such as Nedham's theory of wormatick matter. He then gets down to business in chapter seven where he begins by claiming that "I Am not so religiously sworn to the Philosophy of *Aristotle* and *Galen*, as to take upon me the defence of Elements, Qualities, Temperaments, and Humours" (Castle 1667: 134). However, what he offers is a synthesis.

In the first place, he cites Daniel Sennert's claim that the Chymists are mistaken to reject the first qualities from being causes of diseases and points out that the distempers of hot, cold, moist, dry hinder the performance of duty and office of parts of the body (ibid.: 134–135). He expresses surprise that the chymists should reject the first four qualities when "in their own operations they observe a notable disparity between the effects of a dry, and moist heat: and they employ heat, as the common instrument of almost all their operations" (Castle 1667: 136). Then comes the most interesting claim:

> [b]ut whil'st I assert the Efficiency of the first qualities, in the causing of diseases in the humane Body; I would not be understood, to mean by the word Quality, a Being or Entity distinct from matter or Body: But that I apprehend by hot, cold, moist, and dry, the parts of matter, or Atoms so figured, and moved, as to produce those Effects we call heating, cooling, moistning, and drying.
> (Castle 1667: 136)

Here the first qualities are explicated using the corpuscular principles. Castle is grafting mechanism onto Galenism! And Castle goes on to illustrate the sort of corpuscular explanations of the *primae qualitates* he has in mind. As for heat, its effects "cannot be understood to be done by a bare naked quality, but by certain Atoms which are endued with such a motion, figure, and size, as are fit to penetrate, discuss [dissipate, disperse], dissolve, and perform all those effects which we usually attribute to heat" (Castle 1667: 137). And in the case of cold,

> it is then the property of cold to congeal, fasten, and close together; and those Atoms, which by their shape and figures are fit and proper for those effects, may, with very good Reason, be called atoms of Cold; and Bodies made up of such Particles, cold Bodies.
> (Castle 1667: 137)

Moistness is

> nothing else but a kind of fluidness; and Liquors are commonly said to be moist, inasmuch as when they are poured upon hard and compact bodies, some small parts of them are left behind, either sticking in the

> little Cavities of the Surface, and then the body is said to be wet; or else have insinuated themselves into the most inward pores and recesses of the hard body, which then we commonly say is moistned.
>
> (Castle 1667: 138)

Finally, dryness is "nothing else but a kind of firmness . . . void of all moisture" (ibid.).[14] Castle concludes then,

> I cannot see why these four first Qualities (as they are term'd) should be excluded from having a share in the number of causes of Diseases, since they are notably active (especially the three first) modifications of matter; and not only apt to excite various motions, and cause, as well new Combinations, as dissolutions of bodies in the great World; but also powerfully to alter the Microcosm, and produce sundry different Symptoms, in relation to the motions and harmony of the humane Engin.
>
> (Castle 1667: 138–139)

Thus, Castle retains the central role for the *primae qualitates*. He simply gives them a corpuscular interpretation adverting to the shape, size, and motion of atoms that account for heat and cold, etc. He also retains the theory of humours but his reference to the "humane Engin" shows that this really is a synthesis of the mechanical philosophy with Galenism.[15] It is little wonder then that Castle can assert that

> It is not, I think, to be question'd, that a man is as Mechanically made as a Watch, or any other Automaton; and that his motions, (the regularity of which we call Health) are perform'd by Springs, Wheels, and Engines, not much differing, (except as to the curiousness of their Work) from those pieces of Clock-work, which are to be seen at every Puppet-play.
>
> (Castle 1667: 5–6)

ROBERT SPRACKLING

Robert Sprackling is not well known to historians of the philosophy of science. He has no entry in the new Oxford Dictionary of National Biography. His *Medela Ignorantiæ or A Just and plain Vindication of Hippocrates and Galen from The groundless Imputations of M. N.* of 1665 is the most vitriolic of all the replies to Nedham's book. It is clever but lacks philosophical depth.

Sprackling is a keen defender of Galenism. His main point with regard to the new rival explanatory principles of the chymical and corpuscularian philosophers is that these new principles might be more fundamental than the principles of Galenism, but that this neither implies that the elements, first qualities, and humours need to be abandoned, nor that these new explanatory principles are all they are cracked up to be. Here is his claim *in extenso*:

> The common Doctrine of elements, humors, and temperaments, doth explain the appearances of bodies, and their affections in a gross and general manner, without descending to more particular explications of abstruse and occult causes. The Chymical Philosophy more apertly [plainly] explaineth them, by further dividing composites into their constituent and separable parts; the process whereof the great Physicians of our Age do therefore exceedingly applaud and advance. Yet again it must be acknowledged that this Philosophy is not so perfect or satisfactory to the curious, but that it may be still completed by the Atomical improvers, when they can find any end of their scrutiny and indagations. But in the mean time, it must be granted by all, that these gross notions of humors and temperaments, have served eminently to the Cure of great Diseases for many hundreds of years, and when new curious explications are found (the perfection of Chymistry omitted) they will not alter the Disease or Remedy, but only dilucidate the more secret and recessive natures of both.
> (Sprackling 1665: 45–46)

In short, the Galenic principles are useful enough and might well remain so. As for corpuscularianism, he goes on to say that "(I boldly adventure to affirm) [we] must look on the *Corpuscularian* Philosophy, even in its perfection (if it ever attain it) as an advantage to Ornaments, more than the practice of Physique" (Sprackling 1665: 46). Then toward the end of the book Sprackling deals in more depth with Nedham's "assaulting and storming the whole Fabrique of antient Medicine, but endeavouring to substitute new Principles" (Sprackling 1665: 147). Sprackling claims first that two of the chymists' principles, phlegm and caput mortum, are identical to the Galenists' water and earth. Moreover, the chymists' notions of salt and sulphur were explicated in detail by Hippocrates, Aristotle, and Galen, though admittedly "in passages here and there occasionally disseminated, without such prolix Descriptions" as one finds in the writings of the chymists, such that "the *supposed* new Opinions are only Paraphrases and Enlargements on the old Text, or at the best, experimental Probations." And as such "the later Principles ... serve to elucidate and explicate the natures of the old received Elements, so on this very ground and reason they cannot be justly supposed, to contradict ... or exclude them" (Sprackling 1665: 148–149).

Sprackling is unimpressed by Nedham's appeal to corpuscles, which the latter had labelled Bodikins, and he has another swipe at the notion of indivisible particles while extolling the virtues of observation.

> Not moved by the Notion of his *Bodikins* to this circumspection, but by the surer light of daily experience in their practice, which is not

to be regulated by new, though ingenious, Inventions (little understood, as appeareth by his tearms of *indivisible particles*, a notion salving [explaining] no one appearance of nature and derided by all others that prætend (as he doth) to the new Philosophy and Dissection instead of abstraction of naturals, and yet expressed, at the same instant, when he ridiculously accuseth the learned *Valesius* of want of apt tearms on the same Subject of Atoms) but by serious and judicious observations.

(Sprackling 1665: 64–65)

He then goes on to quote Boyle's claim in the *Usefulness of Natural Philosophy* that

We are yet for ought I can find, far enough, from being able to explicate all the Phænomena of nature, by any Principles whatsoever. And even of the atomical Philosophers, whose Sect seems most ingeniously to have attempted it, some of the eminentest (in all likelyhood then as eagle-sighted as this Machine that wrote *Medela Medicinæ*) *have themselves freely acknowledged to me, their being unable to do it convincingly to others, or so much as satisfactorily to themselves.*

(Sprackling 1665: 65)

The cutting reference to Nedham as an "eagle-sighted Machine" shows that Sprackling was fully apprised of the mechanical philosophy and its approach to nature. Indeed he goes on to mock corpuscular explanations of the structure of Nedham's wormifick matter and refers him to

some young Gentleman of *Cambridge* or *Oxford* that may tutor him in *Gassendus*, and shew him such things as will put him into as high an extasie in seeing things so strange to his knowledge, as the Maid recovered from her native blindness suffered at her first view of the variety of common sensibles.

(Sprackling 1665: 74–75)

The reference to Gassendi brings into play our third and final respondent to Nedham's *Medela medicinae*, John Twysden.

JOHN TWYSDEN

John Twysden (1607–1688) is also a shadowy figure for historians of early modern philosophy and yet his treatment of the principles of the new philosophy is the most philosophically sophisticated and interesting of all of those who took up the cudgels against Nedham. A hint of this is suggested in the very interesting biographical comment that Twysden makes about his relations with Gassendi in his *Medicina veterum vindicata* of 1666. When discussing atomism Twysden mentions Gassendi as

> a learned Philosopher and Divine . . . whom my self had the honour particularly to know, and frequently converse with there [in Paris], and often about this subject; I found him a man very communicable, but to me would never declare his opinion to agree with that of *Epicurus*, onely resolving to write his Life and Philosophie, thought fit to propound fairly what might be said on that subject.
> (Twysden 1666: 138–139)[16]

But before we turn to Twysden's treatment of the various competitors to the Galenic principles, there is a very important passage at the opening of his book that sets out the central thesis of his own medical philosophy. Twysden explains that he is going to reply to Nedham's attack on the Galenic principles, that is "the composition of bodies from the four Elements, the doctrine of Humors, and the combination of Qualities, of which I shall have occasion to speak hereafter" (Twysden 1666: 5). He goes on to state that these principles

> are like those Eccentrick Circles and Epicycles in Astronomy, which are not necessarily true, yet serve to reconcile appearances, and therefore are of equal value, as if they were undoubtedly so; since by them we are brought to a certain knowledge of that truth we seek for, at least such a knowledge as the nature of the thing sought for is capable of.
> (Twysden 1666: 5–6)

Twysden is a self-confessed instrumentalist about the Galenic principles! More importantly, however, it shows that some Galenic physicians, at least, were able to use the traditional principles as a tool without buying into any ontological commitments as to the underlying nature of disease or the composition of bodies and chemical remedies.

Turning next to Twysden's extended discussion of the various principles that were in competition among the physicians, chymists and natural philosophers, we find first a survey of the problem of prime matter and the nature of privation in Aristotle and Plato. Twysden leaves this aside as being "wholly Metaphysical" and turns instead to a critique of atomism (ibid.: 136–147). It is in this context that we find the allusion to his acquaintance with Gassendi. After dismissing atomism he turns to Descartes's theory of matter (ibid.: 147–156) and then the principles of the chymists (ibid.: 156–177) and finally the Aristotelian way, which he finds to be explanatorily superior. The chymists' principles he finds do not replace the Galenical ones and indeed in some cases overlap and in other cases might be deeper explanation of them. One concern is to retain the opposition between qualities because without this bodies would remain inert and unchanging. The discussion throughout is clear, erudite, and balanced. He finds no conceptual or experimental grounds for dismissing the Galenic principles and concludes that the four qualities "have so far prevailed with the world, that they have I think, for nigh, if not full 2000 years, been thought reasonable, and therefore not so easily to be exploded, and thought dry and jejune notions as our *M. N.* would have them." But consistent with his instrumentalism he concludes, "Let every man, however, for me, safely enjoy his own opinion, and the learned judge which carries most weight of reason" (Twysden 1666: 183).[17]

CONCLUSION

It is clear then that the mechanical philosophy, and in particular the new corpuscular principles of shape, size, and motion, did pose a challenge to the foundational principles of Galenic medicine. This chapter has examined just one site among several in which this contest manifested itself. The responses of Galenic physicians to this new philosophy varied greatly. George Castle took a synthetic approach, using corpuscular principles to explicate the first four qualities and so preserving the traditional medical theory. Robert Sprackling dismissed the threat of the new principles arguing that at most they complemented or augmented the traditional ones and that they were far from well tested or perfected. John Twysden took the instrumentalist line, but saw conceptual problems for the main rivals to the elements, qualities, and humours of the Galenists such that the traditional view remained the most efficacious explanatory system. Boyle, for his part, argued against both the Galenists and the chymists that the corpuscular principles were explanatorily superior to any alternative principles on offer, while Nedham's fairly flimsy case against Galenism was left in tatters by his antagonists.

But the very fact that this debate could take place is illustrative of the inroads that the new philosophy was making in medical theory in the mid-seventeenth century. In some respects the Galenic physicians were formidable opponents for the mechanical philosophers, not only because they had so much to lose in terms of credibility and business, but also because of their learning and the many gifted controversialists within their ranks. However, slowly but surely the Galenic principles were being undermined and corpuscular-style explanations in medicine were gaining ground. The theory of matter on which Galenism was built was beginning to vanish.

NOTES

1. See Anstey 2000.
2. See Cook 1986, 1987, 1989; Debus 2001; Wear 2000; Clericuzio 1993.
3. See Frank 1980; Cook 2010.
4. For a recent survey treatment see Cook 2006.

5. Translation by Lawrence M. Principe (= Walmsley and Meyer 2009: 15).
6. Thomson 1665: 65.
7. See also Charleton 1652, which bears the strong imprint of Cartesian thought.
8. The summary "atomic" theory is found on pp. 115–124 of the appended essay to Highmore 1651. For further discussion of the early reception of French corpuscular philosophy and of Highmore see Frank 1980: 92–101.
9. Thomas Willis had appealed to the corpuscular principles in his *Diatribae duae* (1659), but his matter theory was not strictly corpuscularian. He held to the pentad of the chymists. The physician Walter Charleton's *Physiologia Epicuro-Gassendo-Charltoniana*, 1654, espoused an atomist matter theory derived from that of Pierre Gassendi, but I have found no evidence of it having any direct influence on the debates on the nature of matter in the writings of British physicians in the mid-1660s.
10. For Boyle's early interest in medicine see Frank 1980; Hunter 2000b; Kaplan 1993. For Boyle and Starkey see Newman 2003; Newman and Principe 2002.
11. For older studies of Nedham and responses to him see Jones 1961: 206–236 and King 1970: 145–180.
12. See also Nedham 1665: 256.
13. For George Thomson on seminal disease see Thomson 1666: 26–38. For Locke's early seminal theory of disease see "Morbus" in Walmsley 2000. Another version of this Helmontian approach to disease is found in chapter 15 of Thomas Willis' "De febribus" in Willis 1659: 182–200. For background to the seminal theory of disease in this period see Shackelford 1998.
14. Castle accepts that there are humours other than the traditional four, but adheres to the doctrine of temperaments: "men are not improperly said to be of a melancholic, cholerick, or some other temperament; inasmuch, as by how much the more vigorous or remiss the natural heat is in their bowels and entrals, by so much the more weakly or powerfully concoctions are perform'd . . . ," ibid.: 139.
15. Castle (1667: 141) also claims that "I do not find, that the Improvements which have been made in the Theory of Physick, have much altered the Practice."
16. Such a contact with Gassendi is consistent with the interest in the Frenchman's philosophy by a number of English physicians in the early 1650s. Interestingly, earlier in *Medicina veterum vindicata* (Twysden 1666: 117–118) he had claimed of Boyle "I have had the honour to have been particularly acquainted with him now upwards of twenty years" and that "I have frequently been in his Laboratory, seen and been from him made partaker of many of his Preparations . . . [and] have received from his own hand not onely the manner of the Preparations, but the Medicines themselves." Michael Hunter alerted me to the fact that Twysden does appear in Boyle's Workdiary 13 entry #24, made in November 1655 (Royal Society Boyle Papers vol. 25, p. 153), which entry appears to confirm Twysden's claim.
17. Twysden affirms Francis Bacon and the "Experimental Philosophy" as furnishing new truths of nature without disparaging the ancients, Twysden 1666: 102–103.

BIBLIOGRAPHY

Anstey, P. R. (2000) "Descartes' cardiology and its reception in English physiology" in eds. S. W. Gaukroger, J. Schuster, and J. Sutton 2000, pp. 420–444.
Borelli, G. A. (1680–1681) *De motu animalium*, Rome.

Boyle, R. (1999–2000) *Works of Robert Boyle*, 14 vols, eds. M. Hunter and E. B. Davis, London: Pickering and Chatto.
Castle, G. (1667) *The Chymical Galenist*, London.
Charleton, W. (1652) *Darknes of Atheism Dispelled by the Light of Nature*, London.
———. (1654) *Physiologia Epicuro-Gassendo-Charltoniana*, London.
Clericuzio, A. (1993) "From van Helmont to Boyle: A study of the transmission of Helmontian chemical and medical theories in seventeenth-century England," *British Journal for the History of Science*, 26: 303–334.
Cook, H. (1986) *The Decline of the Old Medical Regime in Stuart London*, Ithaca: Cornell University Press.
———. (1987) "The Society of Chemical Physicians, the new philosophy, and the Restoration Court," *Bulletin of the History of Medicine*, 61: 61–77.
———. (1989) "Physicians and the new philosophy: Henry Stubbe and the Virtuosi-Physicians" in eds. R. French and A. Wear 1989, pp. 246–271.
———. (2006) "Medicine" in eds. K. Park and L. Daston 2006, pp. 407–434.
———. (2010) "Victories for Empiricism, failures for theory: medicine and science in the seventeenth century" in eds. C. T. Wolfe and O. Gal 2010, pp. 9–32.
Descartes, R. (1662) *Renatus Des Cartes de homine*, Leiden.
———. (1998) *The World and Other Writings*, trans. and ed. S. W. Gaukroger, Cambridge: Cambridge University Press.
Debus, A. (2001) *Chemistry and Medical Debate: van Helmont to Boerhaave*, Canton, MA: Science History Publications.
Debus, A. and Walton, M. T., eds. (1998) *Reading the Book of Nature: The Other Side of the Scientific Revolution*, Missouri: Sixteenth Century Journal Publishers.
Frank, R. G. Jr. (1980) *Harvey and the Oxford Physiologists: A Study of Scientific Ideas*, Berkeley: University of California Press.
French, R. and Wear, A., eds. (1989) *The Medical Revolution of the Seventeenth Century*, Cambridge: Cambridge University Press.
Gaukroger, S. W., Schuster J. A., and Sutton, J., eds. (2000) *Descartes' Natural Philosophy*, London: Routledge.
Harvey, W. (1651) *Exercitationes de generatione animalium*, London.
Highmore, N. (1651) *The History of Generation*, London.
Hunter, M. (2000a) *Robert Boyle 1627–1691: Scrupulosity and Science*, Woodbridge: Boydell Press.
———. (2000b) "The reluctant philanthropist: Robert Boyle and the 'Communication of Secrets and Receits in Physick'" in Hunter 2000a, pp. 202–222.
———. (2000c) "Boyle versus the Galenists: a suppressed critique of seventeenth-century medical practice and its significance" in Hunter 2000a, pp. 157–201.
Jones, R. F. (1961) *Ancients and Moderns: A Study of the Rise of the Scientific Movement in Seventeenth-Century England*, 2nd edn, St. Louis: Washington University.
Kaplan, B. B. (1993) *"Divulging of Useful Truths in Physick": The Medical Agenda of Robert Boyle*, Baltimore: Johns Hopkins University Press.
King, L. S. (1970) *The Road to Medical Enlightenment 1650–1695*, London: Macdonald.
Le Febvre, N. (1664) *A Compleat Body of Chymistry*, London.
Locke, J. (1975) *An Essay concerning Human Understanding*, 4th edn, ed. P. H. Nidditch, Oxford: Oxford University Press; 1st edn 1690.
Nedham, M. (1665) *Medela medicinae*, London.
Newman, W. R. (2003) *Gehennical Fire: The Lives of George Starkey an American Alchemist in the Scientific Revolution*, Chicago: University of Chicago Press. First published in 1994, Harvard: Harvard University Press.

Newman, W. R. and Principe, L. M. (2002) *Alchemy Tried in the Fire: Starkey, Boyle and the Fate of Helmontian Chymistry*, Chicago: University of Chicago Press.

Park, K. and Daston, L., eds. (2006) *The Cambridge History of Science. Volume 3: Early Modern Science*, Cambridge: Cambridge University Press.

Shackelford, J. (1998) "Seeds with a mechanical purpose: Severinus' semina and seventeenth-century matter theory" in eds. A. Debus and M. T. Walton 1998, pp. 15–44.

Sprackling, R. (1665) *Medela Ignorantiæ*, London.

Thomson, G. (1665) *Galeno-Pale*, London.

———. (1666) *LOIMATOMIA: or the Pest Anatomized*, London.

Twysden, J. (1666) *Medicina veterum vindicata*, London.

Walmsley, J. C. (2000) "'Morbus'—Locke's early essay on disease," *Early Science and Medicine*, 5: 367–393.

Walmsley, J. C. and Meyer, E. (2009) "John Locke's "Respirationis usus": text and translation," *Eighteenth Century Thought*, 4: 1–28.

Wear, A. (2000) *Knowledge and Practice in English Medicine, 1550–1680*, Cambridge: Cambridge University Press.

Willis, T. (1659) *Diatribae duae medico-philosophicae*, London.

Wolfe, C. T. and Gal, O., eds. (2010) *The Body as Object and Instrument of Knowledge: Embodied Empiricism in Early Modern Science*, Dordrecht: Springer.

5 Without God
Gravity as a Relational Quality of Matter in Newton's *Treatise*

Eric Schliesser

In this chapter I interpret Newton's speculative treatment of gravity as a relational, accidental quality of matter that arises through what Newton calls "the shared action" of two bodies. In doing so, I expand and extend on Howard Stein's views (Stein 2002). However, in developing the details of my interpretation I end up disagreeing with Stein's claim that for Newton a single body can generate a gravity/force field.

I argue that when Newton drafted the first edition of the *Principia* in the mid 1680s, he thought that (at least a part of) the cause of gravity is the disposition inherent in any individual body, but that the force of gravity *is* the actualization of that disposition; a necessary condition for the actualization of the disposition is the actual obtaining of a relation between two bodies having the disposition. The cause of gravity is not essential to matter because God could have created matter without that disposition. Nevertheless, at least a part of the cause of gravity inheres in individual bodies and were there one body in the universe it would inhere in that body. On the other hand, the force of gravity is neither essential to matter nor inherent in matter, because (to repeat) it *is* the actualization of a shared disposition. A lone part-less particle would, thus, not generate a gravity field.

Seeing this allows us to helpfully distinguish among a) accepting gravity as causally real; b) positing the cause(s) (e.g. the qualities of matter) of the properties of gravity; c) making claims about the mechanism or medium by which gravity is transmitted. This will help clarify what Newton could have meant when he insisted that gravity is a real force. I present my argument in opposition to Andrew Janiak's influential paper (Janiak 2007).

The view I attribute to Newton is the view that he held when writing the first edition of the *Principia* in the mid 1680s. My evidence for this is a translation of the first draft of Book III of the *Principia* that appeared in English as *A Treatise of the System of the World* (hereafter *Treatise*).[1] I know of no reason to deny its having been written during the mid 1680s, as Newton was moving from the successive drafts deposited with the Royal Society, known as *De Motu*, and the publication of the first edition of the *Principia*. There are two reasons to take it very seriously.

First, it gives us insight into Newton's thinking while he was working out the details of his system. In the published introduction to Book III of the *Principia*, Newton writes that he suppressed his *Treatise* in order to "avoid lengthy disputes" with others' "preconceptions" (Newton 1999: 793). The prejudices he has in mind are not merely "vulgar," but rather philosophical, that is, those in circulation among the learned (especially Hooke and Huygens). For the *Treatise* is more speculatively metaphysical than the published version of Book III. Nevertheless, because of the timing of its writing and the fact that Newton clearly did not disown it, it is rather surprising that it has been largely neglected in Newton scholarship. While many later much-studied additions to the *Principia* can be explained as Newton's response to new empirical evidence or corrections to obvious problems, many of the changes should be viewed as Newton's evolving response to the concerns expressed by some of his religiously motivated interlocutors and to his evolving arguments with Leibniz and his followers.

It is a bit strange that the "General Scholium" and, say, the "Letter to Bentley" have received a lot more attention than the *Treatise* among those who claim to be interested in Newton's *metaphysics*, regardless of how one views the relationship between Newton's technical claims and the material he segregated into queries and the "General Scholium," which was added to the second edition almost certainly to stop readers from alarmed over his theological outlook.[2] Surely in uncovering Newton's substantive commitments, we should not focus primarily on the views he expressed for reasons associated with concerns over his religious, moral, and political views. Elsewhere I call this the "Socratic problem."[3] With Newton's general secrecy and flirtations with Arianism, there is no denying his awareness of such constraints; in his unpublished "De Gravitatione" he calls attention to how Descartes "feared" (Newton 2004: 25) positions that might be thought to offer "a path to atheism" (Newton 2004: 31).

Second, because the *Treatise* was published so shortly after Newton's death it is invaluable in helping us understand the reception of Newton in the eighteenth century. While that reception was shaped by the *Optics*, the Leibniz-Clarke correspondence, and Newtonian expositors, the *Treatise*'s influence is much neglected. This is unfortunate. For example, among Newton scholars it is becoming unfashionable to read Newton as an instrumentalist or positivist, but many probably still read the eighteenth-century British Empiricists as reading Newton in this (mistaken) way and, thus, as uninformative on matters of Newton interpretation (Smith 2001). Moreover, scholars are thus also blind to the fact that British Empiricists are often ambivalent about the new role of Newton's "authority."[4] While here I neither will argue for the importance of the *Treatise* to eighteenth-century readers nor hope to rehabilitate eighteenth-century British thought as a guide to interpreting Newton, it is useful to realize that we do not need to focus exclusively on Huygens, Leibniz, and Kant in offering philosophical insight into Newton's thought. So, I view my project here as making possible the recovery of a well informed eighteenth-century view of Newton—one that has no access to "De Gravitatione" or Newton's alchemical writings, but that can use the *Treatise* to evaluate Newton's original commitments.

Moreover, once we realize that many Empiricists understood Newton as the kind of realist they were not,[5] we can also see that part of the great eighteenth-century debate in philosophy is not between Empiricism and Rationalism, but between those philosophers who believe in creating complete systems from the method of inspecting ideas (for instance Descartes, Spinoza, Locke, and Hume) and those who believe in a more piecemeal mathematical-experimental approach (a tradition initiated by Galileo and Huygens). The Newton of "De Gravitatione," now the focus of so much scholarly attention, who unabashedly offers an analysis of the "exceptionally clear idea of extension" (Newton: 2004: 22 and 27) belongs to the first tradition; the Newton of the *Principia* and the *Treatise* is the champion of the second tradition. (Of course, Newton mentions our "ideas of [God's] attributes," in the "General Scholium" but his argument is not based on an inspection of those ideas.) If I am right about this, we should date "De Gravitatione" before the *Treatise* (a claim that is plausible on other grounds too).

While the view I attribute to Newton should be a privileged one for present purposes, I provide five methodological/historiographic reasons to remain agnostic about how this view should be fully squared with other, potentially competing proposals that Newton entertained on such matters (for example, the role and nature of God or a very subtle ether in supplying the mechanism for attraction), as well as about whether or not Newton offered a stable and consistent position throughout his life. First, Newton's manuscripts reveal a man who was willing to entertain and try out many ideas, although when such views found their way into print without solid empirical evidence they tended to be segregated to "scholia," "queries," "letters," and the work of his various followers. Second, without full consideration of his alchemical, political, and religious views at any moment, I despair of discerning a comprehensive view of Newton's speculative metaphysics even at a single point in time. Someday this may be possible. Third, because of the dangerous political and complicated religious context we cannot always take Newton at face value in speculative matters, especially because in some instances Newton hints at his knowledge of an esoteric/exoteric distinction (Snobelen 1997; 1999). Fourth, despite a few notable and isolated exceptions, we do not have a comprehensive view of how Newton's views evolved across and among many issues. It is dangerous to treat any of Newton's speculative views in isolation, but it is not always clear how to fit these to a larger evolving view. Finally, Newton is an extraordinarily terse writer, and sometimes lesser mortals could have used further clarification. In fact, these considerations have hitherto inclined me to restrict my scholarship to the reception of Newton's views. So despite the firm language in what follows, my views are quite provisional.

Below, I proceed as follows. First I present Newton's relational account of gravity found in his posthumously published *Treatise*. I often contrast my reading with Andrew Janiak's influential essay, especially with his reading of Newton's famous letter to Bentley as ruling out action at a distance. In the final section I disagree with Howard Stein's account of how the gravity field is generated by a particle; I argue that a lone particle is not enough to generate a force field. In offering arguments against Stein's view, I take interpretive stances on the third rule of reasoning and the third law of motion with its corollaries.

GRAVITY AS A RELATIONAL QUALITY OF MATTER

In a recent article, Janiak ably demonstrates that Newton thought that a force of gravity "really exists" (Janiak 2007: 130, 141) because it is one of "the causes which distinguish true motions from relative motions" by way of the "forces impressed upon bodies" (Newton 1999: 412; quoted in Janiak 2007: 134, n.17; see also Janiak 2008: 130–131 and 143–146). Following in the footsteps of Leibniz, Janiak correctly rejects an instrumentalist reading of Newton's views on gravity (Janiak 2007: 138ff).[6]

Because many philosophers are introduced to Newton's views through Clarke's correspondence with Leibniz, and have found Clarke's arguments wanting on a range of issues, they have been prevented from treating Newton as a serious philosopher. There is, thus, no doubt that Janiak's paper and subsequent 2008 book should encourage a re-evaluation of Newton's substantive philosophy on a host of issues.[7] By relying nearly exclusively on Newton's publications and letters Newton wrote to contemporaries, Janiak shows that many of Newton's philosophical views can be gleaned from his published writings if they are read with attention. Newton is often a terse writer, but not obscure. While Newton's unpublished manuscripts on theology and alchemy are fascinating and often provide very helpful context, we need not defer to these to grasp some of the most important strands of Newton's views. This is especially useful if we wish to explore Newton's impact on philosophy, because nearly all these unpublished manuscripts were unavailable to many other influential and insightful readers of Newton in much of the eighteenth and nineteenth centuries (and thereafter).

One of the most important of Janiak's insights is his claim against Leibniz that "if by 'mechanism' one means a natural phenomenon that acts only on the surfaces of other bodies, then Newton rejects the claim that gravity must have some underlying mechanism on the grounds that gravity acts not on the surfaces of bodies, but rather on all the parts of a body (Newton, *Principia*: 943)" (Janiak 2007: 129, n.6). Because "mass is one of the salient variables in the causal chain involving the previously disparate phenomena taken by Newton to be caused by gravity ... [and therefore] *gravity is not a mechanical cause*" (Janiak 2007: 142, 145), Newton rejects a key demand of the mechanical philosophy (Janiak 2007: 146–147).[8]

Nevertheless, in following Leibniz's lead in reading Newton, Janiak ends up misdiagnosing some very important elements of Newton's metaphysics.

In particular, Janiak ignores the *Treatise*. This is important because, without arguing for it, Janiak tacitly attributes to Newton (and the *Principia*) an ahistorical stable position. Focus on the *Treatise* prevents us from ruling out in advance a developmental approach to Newton.[9] In particular, we should be open to allowing Newton to have developed his views between the first and third editions of the *Principia* (and the intervening editions of the *Optics*), although this plays no role in my argument below.

In my interpretation, Newton knows that it is a kind of speculative metaphysics or hypothesis that he deplores with increasing vehemence in others as he anticipates and gets embroiled in debates with the mechanical philosophers and later in vituperative, politicized exchanges with Leibniz and his followers. As Newton becomes ever more insistent on the "empirical" and "experimental" nature of his method (Shapiro 2004), he thus deprives himself of the chance to develop and articulate fully the speculative view that guided the first edition of the *Principia*. Moreover, Newton had a certain amount of self-command in refraining from publicly pursuing certain questions (which we know fascinated him). It follows that he came to devalue the kind of metaphysical interpretation I am about to engage in. Elsewhere I have called attention to what I call "Newton's Challenge," in which the independent authority of philosophy is challenged by empirical science. Newton's intended and unintended role in generating a world in which science and philosophy became competitors is a complex matter, and I cannot do justice to it here (Schliesser forthcoming a). One symptom of this separation of philosophy and science—as Colin MacLaurin, the leading Newtonian of the first half of the eighteenth century, recognized (MacLaurin: 77–79, 96)—is that against systematizing philosophers, Newton and his followers were willing to pursue more narrow research questions. Newton's achievement is to create a mathematical, theoretical structure that was highly promising and efficient as a research engine for ever more informative measurement; this achievement was won at the expense of settling certain metaphysical matters.

In *A Treatise* Newton offers a distinction between a mathematical and natural point of view:[10] "We may consider one body as attracting, another as attracted: But this distinction is more mathematical than natural" (Newton 1969: 37). Here, as in the "technical" contrast between the "mathematical" and "physical" of the *Principia* (Janiak 2007: 131ff),[11] Newton does not deny the reality of the natural point of view. For Newton goes on to write, "the attraction is really common of either [body] to [an]other" (Newton 1969: 37). Newton is alerting his readers to the fact that one cannot simply infer ontology from one's mathematical expression. In the *Treatise's* next paragraph, Newton explains the "natural" perspective more fully, specifically, in terms of the attraction between Jupiter and the sun. I quote two passages.

> There is a double cause of action, to wit, the disposition of each body. The action is likewise twofold in so far as the action is considered as upon two bodies; But as betwixt two bodies it is but one sole single one
> (Newton 1969: 38)

and

> We are to conceive a single action to be exerted betwixt two Planets, arising from the conspiring nature of both.
> (Newton 1969: 39)[12]

In order to get the conception of the nature of an interaction at work in Newton we must note the difference between 1) the "cause of the action," which is "the disposition of each body" and 2) the "action" (or effect) itself. The action is i) twofold as it is *upon* two bodies, and ii) single as *between* two bodies. A way to capture this is to say that a body has two dispositions: a "passive" disposition to respond to impressed forces is codified in the second law of motion, whereas an "active" disposition to produce gravitational force is treated as a distinct interaction codified in the third law of motion.[13]

Thus, we see that the "cause" of the action is "the conspiring nature of both" bodies. For the "conspiring" to occur, the bodies must *share* a "nature" (Newton 1969: 39). To sum up: the cause consists in the "nature" or "disposition" of two bodies (or a twofold cause, because involving two bodies), but it is one interaction or "nature." (For my argument below it is useful that Newton thinks of the interaction itself as a "nature.")[14] What are caused are *one interaction* and *two "actions upon bodies"*; there are two impressed forces. As Howard Stein explains (while using somewhat anachronistic language), "exactly those bodies that are susceptible to the action of a given interaction-field are also the sources of the field."[15]

Newton is offering his readers a radical new idea, one that he feared would encounter a lot of prejudice; here we have on offer a speculative hypothesis about the nature of matter. First, Newton emphasizes a single "action" between two bodies. (He uses the repeated *actio* in Latin.) This is, thus, a very clear description of action at a distance; applying the third law of motion is not merely a mathematical statement, but action at a distance really takes place in nature.[16] In the *Treatise*, Newton explains that this is due to a *shared* quality of two bits of matter. Here Newton offers a hypothesis about the physical cause of the sort he came to reject firmly in the "General Scholium." Somewhat paradoxically, according to Newton the body with a "passive" disposition to be attracted is part of the *cause* of the gravitational force. Newton's genius is that instead of untangling this knot, in the *Principia* he focuses on defining measures of the gravitational force. In the *Treatise* Newton is silent, however, on whether or not this attraction is mediated through a medium.

So, we should distinguish among a) the force of gravity as a real cause (which is calculated as the product of the masses over the distance squared); b) the cause of gravity (which is at least, in part, the masses to be found in each body); b*) "the reason for these [particular] properties of gravity" ("General Scholium," *Principia*, quoted in his own translation by Janiak

2007, 129); and c) the medium, if any, through which it is transmitted.[17] So, the view presented here by Newton takes a complete stance on "a" and a partial stance on "b," but is silent on "b*." In the *Treatise*, Newton is entirely silent on "c," the invisible medium, to explain in what way momentum could be exchanged between two bodies. Given that he uses the language of "action" and is completely silent on the possibility of a medium of transmission, the natural reading of this passage is I) Newton's endorsement of action at a distance with II) the start of an explanation of the cause of gravity in terms of some of the qualities of matter. But instead of being distracted by the incomplete nature of explanation, Newton oriented natural philosophy toward a sophisticated program of piecemeal problem solving through measurement.

Janiak (2007: 137) quotes Leibniz's last letter to Clarke: "For it is a strange fiction . . . to make all matter gravitate, and that toward all other matter, as if all bodies equally attract all other bodies according to their masses and distances, and this by an attraction properly so called . . . , which is not derived from an occult impulse of bodies, whereas the gravity of sensible bodies toward the center of the earth ought to be produced by the motion of some fluid."[18] I agree with Leibniz (*contra* Janiak, 2007: 141) that Newton can be read as endorsing the "strange fiction," and I reject Janiak's apparent willingness to accept Leibniz's insistence (against the fiction) that a motion of some fluid *must* be the "cause" of gravity. Thus, while Janiak criticizes Leibniz's argument against Newton, throughout his article, Janiak conflates i) treating of the mechanism or medium by which gravity is transmitted with ii) treating of the cause of the gravitational qualities of matter (Janiak 2007: 136); while outside the *Principia*, Newton can sometimes be read this way, it appears that Leibniz is the source of Janiak's conflation.

Before I note some obvious objections to my *Treatise*-inspired interpretation of Newton, note three implications of the view I attribute to Newton. First, it treats gravity (understood as a causally real force) as itself "produced" by a relational quality of matter. It is only because of matter's special relationship to other parts of matter that it gravitates. That is, as Smeenk has noted, we "need a pair of bodies for both the 'active' and 'passive' aspects of the disposition to gravitate to be in play; it is only with the pair that the 'active' disposition to produce force combined with the 'passive' disposition to respond causes an interaction that yields actual motions" (Smeenk, personal correspondence, May 23, 2008).

Second, gravity is, thus, not an intrinsic quality of a single bit of matter. The relation only holds between bits of matter, which are said to share a "nature," that is, a "disposition" to gravitate when conjoined.[19] So, in my view gravity is intrinsic to the relation, a "nature," but not to matter itself. For those disinclined *a priori* to believe that Newton would hold such a position, it might help to note that this dispositional, relational account is analogous to a claim made in terms of capacities in "De Gravitatione":

for when the accidents of bodies have been rejected, there remains not extension alone, as [Descartes] supposed, but also the capacities by which they can stimulate perceptions in the mind.
(Newton 2004: 35)

A lone particle in the universe will not stimulate a perception until there is a mind present. (Of course, the official argument goes in the other direction: the description of corporeal nature is "deduced ... from our faculty of moving our bodies!" (Newton 2004: 30)).[20]

Third, because gravity is not an intrinsic quality of a particle of matter, we can grasp why for Newton gravity is not essential to matter. John Henry has argued that this is due to Newton's commitment to God superadding gravity to matter.[21] My view cannot rule out this possibility, and Newton was probably eager to encourage it among certain of his more orthodox interlocutors, such as Bentley. However, I read Newton as claiming in the *Treatise* that gravity is in matter not by, say, God's superaddition, but rather generated by (for lack of a better term) the interaction between at minimum two bits of matter due to their shared nature, that is, mass. Now, as Ori Belkind has reminded me, the shared mass of the two particles is reducible to the distinct masses that each object possesses separately, and this may be thought to make gravity an inherent property; or, given that mass is not a relational quality, one can have the *cause* of gravity defined nonrelationally, and gravity itself defined relationally. (Some of this will be clearer below from my discussion of Rule III and the difference between universal and essential properties of matter.)

But it is only by virtue of the masses entering into an "interaction" that they "produce" gravity. Thus, even after creation, a lonely part-less particle of matter in the universe would not be said to gravitate. The *disposition* is only *actualized* when the universe contains at least two bodies (or a body with parts). So, first, when it comes to gravity, we are dealing with a relational and inessential property of matter. Second, even while we can think of "part" of the "cause" of gravity, mass, as intrinsic to matter, it only "is" a cause when related to other masses.[22]

My reading has seven nice features associated with it. First, while it is a species of speculative metaphysics that Newton increasingly came to deplore, it provides an account of part of the physical cause of gravity in an ontologically sparse way, in accord with Newton's first rule of reasoning.[23] No new entities are introduced. Of course, the downside is that surprising and strange qualities are attributed to known entities. Thus, second, it accords with the ontological priority of matter over the laws of nature in Newton's late cosmological query that:

> it may be also allowed that God is able to create particles of matter of several sizes and figures, and in several proportions to space, and perhaps of different densities and forces, and *thereby* to vary the laws

of nature, and make worlds of several sorts in several parts of the universe.

(Query 31, Newton 1979: 403–404; emphasis added)

It is the relationships among the densities and forces of matter to space that accounts for the varying laws and worlds.

Third, related to these, is an additional attractive feature: it is theologically neutral. Newton leaves room for a possible role for God (for example, as the medium, or as cause of the world), but he is not required to commit to it. Newton writes in a different context in *Optics*, Query 31, it only

> seems probable to [Newton] that God in the beginning formed matter in solid, massy, hard, impenetrable, moveable particles, of such sizes and figures, and with such other properties, and in such proportion to space.
> (Newton 1979: 400)

Fourth, this probabilistic and empirical approach to matters related to our knowledge of God's attributes, which are treated "from phenomena," fits the overall character of the *Treatise*, where God shows up only once (and then very briefly) in the antecedent of a conditional statement about the placement of the planetary orbits (Newton 1969: 33).[24] It even fits the more theologically ambitious picture of "De Gravitatione," where Newton's treatment of the idea of space is compatible with the claim that it is merely a) "as it were an emanative effect of God" (Newton 2004: 21) and b) "an emanative effect of the first existing being" (Newton 2004: 25) as well as c) "the emanative effect of an eternal and immutable being" (Newton 2004: 26); while the reader is invited to think of God as the first cause, all three versions are compatible with a theistic and pantheistic interpretation.[25] Regardless of Newton's private commitments, in public he is nearly always careful not to overstep the evidence.

Fifth, gravity is not essential to matter because not only can we conceive that God would initially create matter without gravity, but also—and more importantly, Newton writes (in a passage also quoted by Janiak) in his comment on the third rule of reasoning:

> I am by no means affirming that gravity is essential to bodies . . . Gravity is diminished as bodies recede from the earth.
> (Newton 1999: 796)[26]

That is, it is an empirical discovery that the strength of gravity can vary with distance. The value of gravitational mass does not vary; what varies is the strength of the interaction. A way to make sense of Newton's denial of having claimed that gravity is essential to bodies is to argue that gravity follows from a relation. Against Janiak I see no textual evidence that Newton ever equated "attraction at a distance" with the claim that gravity is "essential" to bodies (Janiak 2007: 128, n.5).[27] There is no evidence that Newton equated gravity being "essential" to matter with it being "inherent" in matter. To be an "essential" quality means being a quality that is required for (or a necessary condition of) the existence of matter, while gravity being "inherent" says something about its "seat" (Newton 1999: 407), that is, its location. McGuire puts this point nicely: "For Newton, a body can be a body without acting gravitationally" (personal correspondence, May 10, 2008). Of course *were* one to determine that gravity is an essential quality of body, one would also be inclined to claim it inheres in it. But it is true that Newton's third rule also asserts that while gravity is not essential to bodies being bodies—in the sense that it is not a necessary condition for their existence (bodies could have been arranged otherwise)—it is, until we find contrary phenomenon, a *universal,* empirical fact that gravity is not separable from pairs/systems of bodies.

In order to avoid confusion, I am not claiming that Newton argues *from* the universal fact of gravity *to* its being relational quality of matter. These are distinct points. It also means that if I were to wish to argue directly against the superaddition thesis defended by Henry, I would have to challenge his privileging of the exchange with Bentley and the "General Scholium" over other texts. But that is not my aim here. In this chapter, all I wish to show is that Newton's conception while writing the first draft of the *Principia* is theologically more austere than scholars tend to recognize. This conjoined with the fact that my reading fits with a whole range of Newtonian commitments and the fact that Newton made sure that the readers of the *Principia* were familiar with the existence of the *Treatise,* offer some grounds for thinking that I am offering a plausible alternative to Henry's approach.

Sixth, Newton is committed to the position that

> the various phenomena caused by gravity are such that mass and distance are the only salient variables in the causal chain that involves them. We express this precisely through the law of universal gravitation, asserting that gravity is as the masses of the objects in question and is inversely proportional to the square of the distance between them.
>
> (Janiak 2007: 142)

Janiak is surely correct to emphasize that Newton's mathematical account places constraints on what a physical account should look like because our goal is to supplement rather than replace the mathematical account (Janiak 2007: 136). But Janiak ignores that Newton is also in a position to distinguish conceptually between investigating the material cause of gravity and the medium that facilitates the interaction. We can learn things about the nature of matter, mass, without learning anything about the medium. In fact, given that post *Principia* the medium must have negligible mass, one should say that it has an entirely different "nature" than matter.

Seventh, it avoids attributing to Newton the "absurd" *Epicurean* position in which *passive* matter can *act* at a distance; this is clearly rejected in the fourth letter to Bentley (Newton 2004, 102).[28] But it does not follow from this, as Janiak contends, that Newton rules out action at a distance *tout court*. For Newton's position in the letter to Bentley permits us to understand that in the right circumstances matter can be viewed as "active."[29] This is, in fact, what Newton indicates in Query 31, where he contrasts the "passive principle by which the bodies persist in their motion or rest, receive motion in proportion to the force impressing it, and resist as much as they are resisted" with "active principles, such as are the cause of gravity, by which planets and comets keep their motions in their orbits, and bodies acquire great motion in falling; and the cause of fermentation," and so on. Shortly thereafter, he lists "gravity, and that which causes fermentation, and the cohesion of bodies" among the "active principles" (Newton 1979: 400–401). This allows, as he explicitly says in his letter to Bentley, that the attractive "agent be material." I should emphasize that in my position Newton neither asserts that matter is altogether active nor passive.[30] It depends on the way we are conceiving things.

Incidentally, because the laws of motion are also said to be "passive" (and in virtue of the ontological priority of matter over laws, noted in my second point above), I reject as un-Newtonian Fatio's late (1705) speculation that gravity is "an immediate effect of the will of God, and one of the first rules by which he controls the universe ... it is not impossible nor even out of probability, that God, by a law, established that matter attracts itself mutually, with a force proportional to its mass and reciprocal with the square of the distance."[31] If God is involved in accounting for action at a distance, it will be in the way he created matter. Thus, I reject Janiak's claim "Newton considered any non-local action to be simply "inconceivable" (Janiak 2007: 144) on two grounds: Janiak misidentifies what is "inconceivable" according to Newton in the letter to Bentley; and he relies on a distinction between "local" and "distant" action (Janiak 2007: 145) that is not to be found in Newton. Let me turn to details of Janiak's arguments next.

ON THE CONCEIVABILITY OF LOCAL AND DISTANT ACTION

Because my reading of the letter to Bentley conflicts with Janiak's and this ramifies through our approaches to Newton, I quote Janiak (2007) at length before I take up the details of his argument. The passage in question is from the letter to Bentley written in 1693, six years after the *Principia* first appeared:

> It is inconceivable that inanimate brute matter should, without the mediation of something else, which is not material, operate upon and affect other matter without mutual contact ... That gravity should

be innate, inherent, and essential to matter, so that one body may act upon another at a distance through a vacuum, without the mediation of anything else, by and through which their action and force may be conveyed from one to another, is to me so great an absurdity, that I believe no man who has in philosophical matters a competent faculty of thinking can ever fall into it.

After quoting the passage, Janiak appends an interpretative note[32] and claims that Newton, thus, "forcefully" rejects "the very idea of action at a distance" (Janiak 2007: 128). In using this passage to motivate his reading of both Newton's rejection of distant action between parts of matter as well as Newton's search for properties of a medium, Janiak finds himself in excellent company with Maxwell, who also uses it to carefully distinguish between Cotes's attribution of "direct" distant action and Newton's position.[33]

But Janiak (and even Maxwell!) quotes the passage selectively. Janiak leaves out the crucial point of the passage, that is, Newton's denial that he is Epicurean. (John Henry [1999, 2007] has made this point forcefully in his papers.) The full passage reads:

It is inconceivable that inanimate brute matter should, without the mediation of something else, which is not material, operate upon and affect other matter without mutual contact, as it must be if gravitation in the sense of Epicurus, be essential and inherent in it. And this is one reason why I desired you would not ascribe innate gravity to me. That gravity should be innate, inherent, and essential to matter, so that one body may act upon another at a distance through a vacuum without the mediation of anything else, by and through which their action and force may be conveyed from one to another, is to me so great an absurdity, that I believe no man who has in philosophical matters a competent faculty of thinking can ever fall into it. Gravity must be caused by an agent acting constantly according to certain laws; but whether this agent be material or immaterial, I have left to the consideration of my reader.
(Newton 2004: 102)

By contrast, my preferred reading of the letter to Bentley is that it is "inconceivable" that "inanimate brute" matter can produce action at a distance (especially) if we conceive gravity along Epicurean lines, that is, as innate, essential, and inherent to matter. But this entirely allows other conceptions of matter with more "active" properties (and other conceptions of gravity). The last sentence of the letter to Bentley quoted above qualifies the reading of absolutist-denial of action-at-a-distance offered by Maxwell and Janiak: Newton says that he will leave it to the reader to decide if gravity is "caused" by a "material or immaterial" "agent"! This means that Newton does not rule out the existence of (properly reconceived) *matter* as an *active* agent or cause of gravity. It would, of course, be a contradiction in terms for "passive" matter to be an "agent"; but Newton never claims in his own voice that matter must always be passive. (Recall my treatment of active/passive principles above.) Newton and later Cotes did consistently deny that "mere mechanical" causes or "necessity of nature"(Newton 1999: 397) or "blind fate" (Newton 2004: 137) can explain the universe. He wants to avoid being read as an Epicurean or Hobbesian. It is surprising that Janiak misses this concern. For, in the fifth reply to Leibniz, which plays a crucial role in Janiak's argument, Clarke forcefully rejects Leibniz's attempts to tag Newton as an Epicurean (§128–130: LC: 96). Of course, everything I say here can be embraced by somebody defending the superaddition thesis.

Unfortunately, Janiak uses his reading of the letter to Bentley to attribute to Newton the view that action at a distance is inconceivable and, thus, that Newton relies on a (tacit) distinction between (inconceivable) distant and (conceivable) local action (e.g. "all action between material bodies must be *local*—on pain of there being an "inconceivable" distant action," Janiak 2007: 143 and 141). I find no evidence in Newton for such a distinction between local and distant action.[34] To be clear, in his analysis of "local" action, Janiak (2007: 143) is careful to distinguish Leibniz's restrictive position, which only permits *surface* action (any causal interaction that involves the surfaces of two or more bodies), from a less stringent version (gravity may involve some kind of medium that does not act on the surface of bodies, but somehow penetrates them) that Janiak attributes to Newton. In his paper Janiak correctly cites Newton's Query 28 as rejecting the Leibnizian hypothesis (Janiak 2007: 143). Janiak also appears to cite Newton's Query 21 to illustrate his claim that Newton is tacitly relying on a local/distant action distinction. But unfortunately, most such claims by Janiak presuppose his further claim that Newton thinks distant action is inconceivable.[35] Janiak ignores very clear evidence to the contrary from the *Principia*. For in the scholium to proposition 69 of Book I of the *Principia*, Newton lists as the very first of four possible, physical explanations of "attraction" that it may be "a result of the action of the [distant!] bodies either drawn toward one another" (Newton 1999: 588). If Newton had thought it inconceivable it would be peculiar for him to leave it in the *Principia* after his exchanges with Bentley.

Equally unpromising is the way that, from Newton's explicit denial of gravity being essential to matter, Janiak moves to attributing to Newton a denial of action at a distance. As noted above, to deny that gravity is essential to matter is just to deny that gravity is a necessary condition for the existence of a lone, part-less particle, or one of the primary qualities of matter. But I have argued against such a view: Newton is committed to gravity being a relational quality.

ON THE NATURE OF THE FORCE[36]

My position does come into conflict with an alternative, influential reading of Newton. Howard Stein's "field" interpretation of gravity was initially and influentially offered as an interpretation of Newton's example in the treatment of Definition 8 of the *Principia,* where Newton writes that "accelerative force [may be referred to], the place of the body as a certain efficacy diffused from the center through each of the surrounding places in order to move the bodies that are in those places; and the absolute force [may be referred], to the center as having some cause without which the motive forces are not propagated through the surrounding regions, whether this cause is some central body (such as the lodestone in the center of a magnetic force or the earth in the center of a force that produces gravity) or whether it is some other cause which is not apparent" (Newton 1999: 407–408). Note, by the way, that Newton distinguishes the cause of the force and it in turn from gravity, offering independent support for my claim above not to conflate such matters. Given that Stein was the original source of my treatment of the *Treatise* passage as authoritative, I conclude by briefly distinguishing my view from Stein's.

Stein treats Definition 8 as claiming that a single particle generates a force field in the places around it.[37] But this does injustice to the fact that according to Newton it is the "shared nature" that generates the inverse-square force. So, on my view it is *only* a *pair of bodies* that generates anything. My dispute with Stein turns on how to interpret Newton's empirical approach, especially in terms of i) the exact wording of the explanation of Definition 8; ii) the formulation in the third rule of reasoning; and iii) perhaps on one's sense of Newton's aesthetics. Let me explain these three differences, and offer responses to Stein's arguments.

In arguing for his field interpretation, Stein is committed to the claim that bodies generate a gravity field around them even in places where there are no other bodies. My problem with Stein's reading is that the use of "field" can suggest to some readers an ontology in conflict with the interpretation that I have been developing. Against Stein, I argue, first, that in Definition 8, Newton does not claim that there would be an accelerative force or an accelerative measure of a gravitational force in the absence of a second body; rather, the definition of accelerative forces is given in terms of the disposition of a central body to "move the *bodies* that are in" the surrounding places. Thus, the example following Definition 8 is quite clear that we are dealing with an interaction (or shared action). The example provides, as Ori Belkind writes, "evidence for the fact that Newton distinguished between the cause of gravity (the masses of bodies) and the force of gravity [in] the distinction between the absolute, accelerative and motive measures of the force. The absolute measure of the force is associated with the cause of the force, located inherently in the body at the center of attraction. The accelerative and motive measures can be relational" (Belkind, private correspondence, May 4, 2008). Moreover, the whole example of Definition 8 is about weight; that is, Newton is giving a treatment of forces not in terms of their (counterfactual) impact on empty places but on *places filled with matter.* This fits nicely with my emphasis above on the ontological priority of matter.

Second, on my reading of the third Rule of Reasoning, we can infer universally that pairs or systems of bodies, or shared actions, generate forces where bodies are, but no more.[38] We have no empirical evidence whatsoever for how to think about a lone part-less particle. In fact, Rule 3 and Newton's explication of it is articulated in terms of the plural "bodies" and their plural "parts." (Recall: "Those qualities of bodies that cannot be intended and remitted and that belong to all bodies on which experiments can be made should be taken as qualities of all bodies universally" (Newton 1999: 795).) The only lone body mentioned in the long discussion accompanying Rule 3 is a hypothetical disconfirming divisible body, but this is unrelated to the universal nature of gravity and is thus evidence for my approach.

Finally, my whole approach makes sense of the fact that Newton's earliest and most able readers, including Leibniz, found even the cleansed view of the *Principia* unabashedly occult. A reading of Newton that makes him out to appear plausibly as a certain kind of innovative scholastic should, thus, have something to be said for it, especially because we are also trying to explain why he would have suppressed it in light of expected prejudices. It is to be admitted that in my view of Newton the infinite universe is populated by infinite number of pairs or systems of interactions within and among bodies generating infinite numbers of attractions. But that is no stranger than the mere fact of universal attraction. Newton is leading us to this view when he writes in his account of the third rule of reasoning:

> if it is universally established by experiments and astronomical observations that all bodies on or near the earth are heavy toward the earth, and do so in proportion to the quantity of matter in each body, and that the moon is heavy toward the earth in proportion to the quantity of its matter, and that our sea in turn is heavy towards the moon, and that all planets are heavy toward one another, and that there is a similar heaviness of comets toward the sun, it will have to be concluded by this third rule that all bodies gravitate toward one another.
> (Newton 1999: 796)

So, on balance, my position should be favored over Stein's because it avoids anachronism and does justice to the text of Newton and the expected reactions to it by his contemporaries. It does saddle Newton with a potentially unbalanced mixture of speculative hypothesis and strict verificationism, and this may be thought to count against the view. (But this offers no solace to Janiak, because he involves Newton in more substantial and less empirically grounded speculative hypotheses.)

CONCLUSION

Let me end with two disconnected observations. In my treatment I have said nothing about a very important and much neglected passage (Book 1, Section 11, Scholium): "finally, it will be possible to argue more securely *concerning the physical species, physical causes, and physical proportions of these forces*" (Newton 1999: 589, emphasis added). By talking about the conspiring nature of matter, my treatment of the *Treatise* has at least started an interpretation of what Newton thought the physical causes were and what species they belonged to. But this is only a tentative start because I have said nothing about Newton's concept of a "natural power."

Finally, I am aware that my somewhat Talmudic reading of Newton is not sexy; other commentators will note three omissions: I neither connect my account of Newton's views to other streams of intellectual thought; nor do I provide Newton with a view that is easily assimilated in our contemporary debates; nor, finally, is my Newton a thoroughgoing theist or Platonizing mystic. But while unfashionable, my Newton is a refined and subtle metaphysician.[39]

NOTES

1. This is a translation of *De Motu Corporum liber secundus*, which appeared as *De Mundi Systemate Liber*, London, 1728 (revised edition 1731), reproduced in Newton 1969.
2. This is a controversial claim. I argue for it in Schliesser (ms).
3. The "Socratic problem" is about how social (religious, political, moral) forces can threaten the independence and authority of philosophy. I distinguish among at least five distinct theses: SP1: a philosopher claims that "practical" philosophy takes precedence (in some way) over "theoretical" philosophy; SP2: a philosopher explains how statements of traditional religious/political texts can be understood as expressions of his/her philosophical doctrines; SP3: a philosopher appeals to non-philosophical (political, religious, social) authorities as justification of doctrine(s); SP4: a philosopher is forced by outside authorities to adjust views; SP5: a philosopher is held accountable for the impact of teachings on students. For detailed discussion, see Schliesser (forthcoming a).
4. See Schliesser 2005, 2008.
5. For example, David Hume claimed that "[i]t was never the meaning of Sir ISAAC NEWTON to rob second causes of all force or energy; though some of his followers have endeavoured to establish that theory upon his authority" (EHU 7.1.25, footnote 16; Hume 2000: 58): a note on Hume's terminology: God is the "first cause;" "second causes" are ordinary finite causes that operate in nature, e.g. laws of nature, or certain species of powers).
6. Janiak (Newton 2004: 134) quotes Clarke's fifth letter to Leibniz as follows: "by that term [attraction] we do not mean to express the cause of bodies tending toward each other, but barely the effect, or the phenomenon it self, and the laws or proportions of that tendency discovered by experience; whatever be or be not the cause of it." In note 18, Janiak refers to sections 110–116 of Clarke's fifth letter in Leibniz, G VII 437–438. The instrumentalism/realism vocabulary does not do justice to the fact that prior to the *Principia*, Newton seems to have allowed that existence is not univocal; as he writes in "De Gravitatione," things can have their "own manner of existing which is proper" to them (Newton 2004: 21). Strictly speaking the quote is only about extension, but context makes clear the doctrine is also applicable to substances and accidents. It is not an isolated occurrence in "De Gravitatione"; Newton also claims: "[w]hatever has more reality in one space than in another space belongs to body rather than to space" (Newton 2004: 27).
7. For the sake of argument, I assume that "philosophical" need not be defined.

8. It is fruitful to understand Newton as redefining the mechanical philosophy.
9. For a developmental view see, for example, Cohen's "A Guide to Newton's *Principia*" in Newton 1999 and McGuire 1970a.
10. My discussion of this material is indebted to Howard Stein's translation and interpretation, shared with me in private correspondence. See also Stein 2002: 287–289.
11. This is often ignored by those that use a line from *Principia*, Definition VIII, "For I here design only to give a mathematical notion of those forces, without considering their physical causes . . . ," to offer instrumentalist readings of Newton. Due to space constraints here, I take no stand on Janiak's and George Smith's important debate over the exact nature of the contrast between the mathematical and physical. See Janiak 2007: 145 and Smith 2002: 150–151.
12. The Latin is: "Causa actionis gemina est, nimirum dispositio utriusque corporis; actio item gemina quatenus in bina corpora: at quatenus inter bina corpora simplex est & unica," Newton 1728: 25 and: "Ad hunc modum concipe simplicem exerceri enter binos Planetas ab utriusque conspirante natura oriundam operationem," Newton 1728: 26.
13. Stein 2002: 289, discussing Query 31 of the *Optics*. Of course, this Query was written much later, and it is not impossible that Newton is extending or developing an original position rather than just articulating it. See also McGuire 1970b.
14. For useful context on Newton's "Platonic" distinction between "being a nature" and "having a nature," see McGuire 2007.
15. See Stein 2002: 288. I argue against Stein's "field" interpretation in the final section.
16. Janiak would agree with me that this reflects "Newton's contention in the *Principia*'s 'General Scholium' that *gravitas revera existat*." But as George Smith first pointed out to me, Janiak's translation of that phrase is a bit misleading because the whole sentence is more plausibly read as a programmatic than an indicative statement.
17. In the body of the text, I distinguish between talking of "qualities" of matter and "properties" of gravity. I believe Newton and Locke use "quality" to pick out a property that can be causally efficacious on minds, a subset of "properties" more generally. I thank Lex Newman for sharing unpublished material on Locke.
18. Janiak cites Leibniz, G VII 397–398.
19. In order to avoid confusion: Newton's dispositional language is used here not to advance an instrumentalist reading (e.g., McMullin 2001 discussed in Janiak 2007: 135).
20. Peter Anstey and Dana Jalobeanu correctly pointed out that there are differences between the matter theory of "De Gravitatione" and *Principia/Treatise*. Fair enough. Here I am using "De Gravitatione" as mild support for the plausibility that Newton was attracted [sic] to a relational view; I am not offering it as an illustration of the view developed in body of text. For an excellent paper on Newton's matter theory and the changes between "De Gravitatione" and *Principia*, see Biener and Smeenk forthcoming.
21. See Henry 1999 and 2007. For a very useful account of superaddition, see Downing 2007: 364ff.
22. One might worry that Newton's second letter to Bentley might cause a problem for my reading. Newton writes, "You sometimes speak of gravity as essential and inherent to matter. Pray do not ascribe that notion to me; for the cause of gravity is what I do not pretend to know, and therefore would take more time to consider of it" (Newton, 2004: 100; I thank Janiak for calling my attention to this passage). While this is evidence for my reading

that Newton denies that gravity is essential and inherent to matter, it also appears to claim that Newton knows nothing about the cause of gravity. But Newton's "therefore would take more time to consider of it" suggests, at least, perhaps, he thought the subject was within his apprehension and worth inquiring about further (i.e. that the cause might be discovered). By my lights Newton should have said he does not know the full cause" so either Newton was not careful or there is a genuine shift from the *Treatise* (perhaps due to considerations of "audience"). I thank Sarah Brouillette for discussion.

23. "No more causes of natural things should be admitted than are both true and sufficient to explain their phenomena" (Newton 1999: 794).
24. Of course, given that it "might be read by many," the *Treatise* itself is not immune from concerns regarding the "Socratic Problem." In the published version of the first edition of Book 3 of the *Principia*, Newton firms up the claim in proposition 8, corollary 5 (Newton 1999: 814).
25. For more on my views on Newtonian emanation, see Schliesser forthcoming c.
26. Added to the third edition of the *Principia*.
27. Nevertheless, although on different grounds, Janiak and I agree in rejecting the so-called Cotes/Kant reading—gravity is an essential quality of matter—adopted in the Editor's preface of the second edition of the *Principia* (Janiak 2007: 143 n.41).
28. Henry 1999 and 2007 make this point forcefully. In correspondence (May 11, 2008), Katherine Brading offered the following useful observation: "If you put a lone Epicurean atom into the void, it will 'know' which way is 'down' and start moving in that direction. If you put a Newtonian [part-less] body into the void it won't 'know' that it's a gravitational body, or 'know' which way to move gravitationally, until you put in a second [part-less] body, and thereby bring the gravitational relation into being."
29. See Joy 2006 for the importance of Newton's distinction between active and passive principles.
30. While this may sound strange, it is by no means unique in Newton. See, for example, Newton's treatment of the "inherent force" of inertia. Newton claims that this force can sometimes be viewed "passively": "Inherent force of matter is the power of resisting"; but sometimes it is more "active": "a body exerts this force ... during a change of its state, caused by another force impressed upon it" (quoted from the third definition). This is why Bertoloni Meli 2006 introduces the language of "potential" or "latent" force to treat the "passive" mode. He captures insightfully the "twofold perspective" (Bertoloni Meli 2006: 325) with which Newton analyses such matters.
31. See the 1949 Gagnebin edition of Fatio; cited in Stein 1970.
32. Janiak's fifth note reads: "See H. W. Turnbull, et al., ed., *The Correspondence of Isaac Newton* [*Correspondence*] (Cambridge: Cambridge University Press, 1959–), vol. 3, 253–254. There is important evidence indicating that this famous passage was not merely the expression of a privately held view, or a public view that Newton later relinquished. The contention that gravity is 'essential' to—or 'inherent' in—matter is construed here to mean that material objects attract one another at a distance. Newton apparently thinks that if gravity is an essential property of material bodies, two spatially separated and otherwise lonely bodies would each bear the property (i.e., they would bear it in a world empty of other bodies). That entails, in turn, that two otherwise lonely bodies would bear the property *in the absence of a medium*. And, of course, the view that spatially separated bodies could interact gravitationally in the absence of any medium is the view that bodies can act on one another at a *distance*. This identification of the claim that gravity is 'essential' to matter with the assertion of distant action is particularly salient because, in an important passage in the *Principia*, Newton writes: 'I am by no means affirming that gravity is essential

to bodies . . . Gravity is diminished as bodies recede from the earth' (Newton 1999: 796). This passage appears in the last (1726) edition of the text. Since Newton never accepted the view that gravity is 'essential' to matter, he apparently never relinquished the contention that distant action between material bodies is impossible."

33. See Maxwell 1890, 2: 316 & 487. Maxwell appears to be influenced by Colin MacLaurin's account. We can understand Smith's "History of Astronomy" (Smith 1795: 29–93) as an attempted correction to MacLaurin. For more on these matters, see van Lunteren 1991.
34. Janiak's treatment of Newton on "local action," has been developed by Kochiras 2009 and challenged in Schliesser (forthcoming b).
35. One might also be tempted to point to a passage from Newton's anonymously published "Account of the *Commercium Epistolicum*," where Newton rejects Leibniz's view of God as "an intelligence above the bounds of the world; whence it seems to follow that he cannot do anything within the bounds of the world, unless by an incredible miracle" (Newton 2004: 125). Now it is true that Newton has just affirmed that God is omnipresent. So, a natural reading of the "incredible miracle" that Newton attributes to Leibniz might be that God acts at a distance (for an omnipresent God can always—in Janiak's terminology—act locally). The passage would then be very ironic: Leibniz accuses Newton of being committed to action at a distance, and Newton turns the table on Leibniz. But this misunderstands Newton's point here; if (Leibniz's) God is above the bounds of the world, this means he is *outside* of space and time altogether. So, it would not be appropriate at all to say that God is acting at a distance; God would literally be *acting* from *nowhere*, that is, an *incredible* miracle. In "De Gravitatione" Newton had claimed "no being exists or can exist which is not related to space in some way" (Newton 2004: 25). Thus, in his response to Leibniz, Newton is echoing the doctrine of the sixth chapter of Spinoza's *Theological-Political Treatise*, where in his discussion of miracles Spinoza rejects the very intelligibility of placing God above the bounds of the world.
36. The discussion below has benefitted from private correspondence with Howard Stein, although the reader should be aware that he objects to my characterization of his position.
37. See Stein 1970. Reading Belkind 2007 made me start to question Stein's use of Definition 8.
38. I came to this conclusion before reading Miller 2009 in manuscript. The earliest draft of Rule 3 reads "The laws of all bodies in which experiments can be made are the laws of bodies universally" quoted in McGuire (1970a: 15). Miller's position and the view presented in the body of this chapter also fit nicely with Brading (forthcoming).
39. I have received detailed comments on earlier drafts of (parts) of this chapter by Ori Belkind, Zvi Biener, Sarah Brouillette, John Henry, David M. Miller, Lex Newman, Chris Smeenk, and Howard Stein. I am also grateful to Ted McGuire for helpful suggestions, and to audiences in Dublin, Oxford, and Gent, including Dana Jalobeanu, Maarten Van Dyck, Karin Verelst, and Carla Rita Palmerino, to which I presented on an earlier drafts of some of this material. Moreover, Katherine Brading helped me formulate my thesis and gave penetrating comments on the whole of the chapter. I am also grateful to Bill Harper, who, despite misgivings about the position, encouraged me to pursue this line of thought after an extensive discussion in an airport lounge. Special thanks due to the editors of this volume for their many insightful comments. I also thank the anonymous referees of *Journal of the History of Philosophy*, who provided helpful comments on

a paper that includes substantial material discussed here. Finally, special thanks are due to Andrew Janiak, who not only gave detailed comments on this chapter, but thanks me (among others) in the final footnote to the paper I criticize here. Ever since Michael Friedman chose to deflect one of my questions to Janiak at a symposium at New York University (November, 2006), my views have been developed in dialogue and correspondence with Janiak.

BIBLIOGRAPHY

Belkind, O. (2007) "Newton's conceptual argument for absolute space," *International Studies in the Philosophy of Science*, 21: 271–293.

Bertoloni Meli, D. (2006) "Inherent and centrifugal forces in Newton," *Archive for History of Exact Sciences*, 60: 319–335.

Biener, Z. and Smeenk, C. (forthcoming) "Cotes' queries: Newton's Empiricism and conceptions of matter" in eds. A. Janiak and E. Schliesser forthcoming.

Brading, K. (forthcoming) "Newton's law-constitutive approach to bodies: a response to Descartes" in eds. A. Janiak and E. Schliesser forthcoming.

Cohen, I. B. and Smith, G. E., eds. (2002) *The Cambridge Companion to Newton*, Cambridge: Cambridge University Press.

Downing, L. (2007) "Locke's ontology" in ed. L. Newman 2007, pp. 352–380.

Henry, J. (1999) "Isaac Newton and the problem of action at a distance" *Krisis*, 8–9: 30–46.

———. (2007) "Isaac Newton y el problema de la acción a distancia," *Estudios de Filosofía*, 35: 189–226.

Hume, D. (2000) *An Enquiry concerning Human Understanding*, ed. T. L. Beauchamp, Oxford: Clarendon Press; 1st edn 1748.

Janiak, A. (2007) "Newton and the reality of force," *Journal of the History of Philosophy*, 45: 127–147.

———. (2008) *Newton as Philosopher*, Cambridge: Cambridge University Press.

Janiak, A. and Schliesser, E., eds. (forthcoming) *Interpreting Newton: Critical Essays*, Cambridge: Cambridge University Press.

Joy, L. (2006) "Scientific explanation: from formal causes to laws of nature" in eds. K. Park and L. Daston 2006, pp. 70–105.

Kochiras, H. (2009) "Gravity and Newton's substance counting problem," *Studies in History and Philosophy of Science*, 40: 267–280.

MacLaurin, C. (1748) *An Account of Sir Isaac Newton's Philosophical Discoveries*, London.

Maxwell, J. C. (1890) *The Scientific Papers of James Clerk Maxwell*, 2 vols, ed. W. D. Niven, Cambridge: Cambridge University Press.

McGuire, J. E. (1970a) "Atoms and the 'Analogy of Nature': Newton's third rule of philosophizing," *Studies in History and Philosophy of Science*, 1: 3–58.

———. (1970b) "Newton's 'Principles of Philosophy': an intended Preface for the 1704 *Opticks* and a related draft fragment," *British Journal for the History of Science*, 5: 178–186.

———. (2007) "A Dialogue with Descartes: Newton's ontology of true and immutable natures," *Journal of the History of Philosophy*, 45: 103–125.

McMullin, E. (2001) "The impact of Newton's *Principia* on the philosophy of science," *Philosophy of Science*, 68: 279–310.

Miller, D. M. (2009) "Qualities, properties, and laws in Newton's method of induction," *Philosophy of Science*, 76: 1052–1063.

Newman, L., ed. (2007) *The Cambridge Companion to Locke's Essay Concerning Human Understanding*, Cambridge: Cambridge University Press.

Newton, I. (1728) *De mundi systemate liber*, London: Tonson. Accessed at: http://books.google.com/books?id=e44_AAAAcAAJ&printsec=frontcover&hl=nl&source=gbs_v2_summary_r&cad=0#v=onepage&q&f=false

———. (1969) *The System of the World*, ed. I. B. Cohen, London: Dawsons of Pall Mall.

———. (1979) *Opticks: Or a Treatise of the Reflections Inflections and Colours of Light*, New York: Dover Publications.

———. (1999) *The Principia: Mathematical Principles of Natural Philosophy*, eds. and trans. I. B. Cohen and A. Whitman, Berkeley: University of California Press; 1st edn 1687.

———. (2004) *Isaac Newton: Philosophical Writings*, ed. A. Janiak, Cambridge: Cambridge University Press.

Park, K. and Daston, L., eds. (2006) *The Cambridge History of Science, Volume 3: Early Modern Science*, Cambridge University Press.

Schliesser, E. (2005) "On the origin of modern naturalism: the significance of Berkeley's response to a Newtonian indispensability argument," *Philosophica*, 76: 45–66.

———. (2008) "Hume's Newtonianism and Anti-Newtonianism," Stanford Encyclopedia of Philosophy (Winter 2008 Edition), ed. E. N. Zalta, http://plato.stanford.edu/archives/win2008/entries/hume-newton/.

———. (forthcoming a) "The Newtonian refutation of Spinoza: Newton's challenge and the Socratic Problem" in eds. A. Janiak and E. Schliesser forthcoming.

———. (forthcoming b) "Newton's substance monism, distant action, and the nature of Newton's Empiricism: discussion of Kochiras," *Studies in History and Philosophy of Science*.

———. (forthcoming c) "Emanative causation, Spinozism, and ontology in Newton," *Foundations of Science*.

———. (ms) "How Epicurean was the first edition of the *Principia*?"

Shapiro, A. E. (2004) "Newton's experimental philosophy," *Early Science and Medicine*, 9: 185–217.

Smith, A. (1795) *Essays on Philosophical Subjects*, London.

Smith, G. E. (2001) "Comments on Ernan McMullin's 'The impact of Newton's *Principia* on the philosophy of science,'" *Philosophy of Science*, 68: 327–338.

———. (2002) "Newton's Methodology" in eds. I. B. Cohen and G. E. Smith 2002, pp. 138–173.

Snobelen, S. D. (1997) "Caution, conscience and the Newtonian reformation: the public and private heresies of Newton, Clarke and Whiston," *Enlightenment and Dissent*, 16: 151–184.

———. (1999) "Isaac Newton, Heretic: the strategies of a Nicodemite," *British Journal for the History of Science*, 32: 381–419.

Spinoza, B. (2007) *Theological-Political Treatise*, ed. and trans. J. Israel, trans. M. Silverthorne, Cambridge: Cambridge University Press.

Stein, H. (1970) "On the notion of field in Newton, Maxwell, and beyond" in ed. R. H. Stuewer 1970, pp. 264–287.

———. (2002) "Newton's Metaphysics" in eds. I. B. Cohen and G. E. Smith 2002, pp. 256–328.

Stuewer, R. H., ed. (1970) *Historical and Philosophical Perspectives of Science*, New York: Gordon and Breach.

van Lunteren, F. (1991) "Framing hypotheses: conceptions of gravity in the 18th and 19th centuries," PhD dissertation, Utrecht University.

Part III
Matter and the Laws of Motion

6 The Cartesians of the Royal Society
The Debate Over Collisions and the Nature of Body (1668–1670)

Dana Jalobeanu

In the first years after its formation, the Royal Society was engaged in a complex process of shaping and presenting a public image of natural philosophy established upon Baconian foundations. Although inside the Royal Society there were different interpretations concerning the exact meaning of "Baconianism,"[1] in the public statements there was a considerable agreement upon what the new philosophy of the virtuosi was supposed to be: a gathering of "uncorrupted eyes" and industrious hands (Sprat 1667: 72),[2] a communal enterprise for gathering together natural histories, witnessing and classifying the results of experiments.[3] Such was the experimental and Baconian make-up of the Society that some members insisted on placing a formal ban upon owning hypotheses or, worse, doctrines.[4]

In many ways the ban on hypotheses is reflected in the structure and methodology of the Royal Society: producing science is seen as a bottom-up process, starting with a careful gathering of natural histories in such a way that observations and experiments are "free from theorizing."[5] Formulating causal connections and theories would be a second layer of works, sometimes relegated either to future ages or to particular fellows of the Royal Society in a way reminiscent of Bacon's scientific utopia.[6] In a similar "Baconian" fashion, the methodological discourse of the Royal Society's apologists is carefully wrapped in moral and religious language. In his *History of the Royal Society*, Thomas Sprat goes to great lengths to emphasize the moral and religious achievements entailed by the proper way of doing experimental philosophy.[7] Meanwhile, as Sprat himself acknowledges, there were other, more practical reasons for banning hypotheses from the new experimental philosophy. One of them, arguably the most important, concerned one of the key issues of the "new" philosophy: the danger of sectarianism (Sprat 1667: 103–104). Adopting a doctrine/theory was seen as the first step toward adhering to a school or sect of philosophers or, worse, attempting to become a founder of a new sect (Jalobeanu 2006a).

That the early Royal Society had to fight sectarian divisions is beyond doubt; what is less clear is how much such general and methodological concerns affected the development of particular philosophical or "scientific" issues and debates. My purpose in this chapter is to show how much insight one can gain in studying such particular examples just by taking seriously the concerns over sectarianism within the early Royal Society: more precisely, by taking seriously the often reiterated assertion stating that in its first years the Royal Society was divided between "Cartesians" and "Gassendists;" so divided, in fact, that the formal ban on hypotheses had to be reiterated as a reminder of the methodological coherence or unity of the society.

Probably the first and best-known form of such a claim is to be found in a book published in 1664, *Relation d'un Voyage en Angleterre*.[8] Samuel Sorbière, royal historiographer to the king of France, an enthusiast of the new philosophy, translator, and popularizer of science, someone "acquainted with the great and learned men in France, Italy, England and Germany," visited England with the declared purpose to be "further instructed into the matters of Literature and the Science." In order to do so, Sorbière became acquainted with as many virtuosi as possible, attended a couple of meetings of the Royal Society and, in his book, painted a vivid and enthusiastic picture of the new scientific body as constructed upon Baconian foundations but subsequently divided in two sects:

> it cannot be discerned that any Authority prevails here; and whereas those who are meer Mathematicians favour *Descartes* more than *Gassendus*, the *Literati*, on the other Side, are more inclined to the latter.
>
> (Sorbière 1709: 38)

In London, the publication of Sorbière's book was a scandal.[9] Thomas Sprat left aside the writing of his *History* just to respond to the unflattering picture painted by Sorbière of English virtuosi, English science, and the English in general. It is worth noting that, while answering the precise remark concerning the camps within the Royal Society, Sprat does not have much to say. He simply emphasizes the new scientific methodology of the virtuosi whose relation to dogmatic philosophy is not the traditional one *and* claims that the mathematicians and the literati live in peace with each other.[10]

Is this exchange of pamphlets between two popular writers to be taken seriously? Do Sorbière and Sprat say something relevant about the constitution of the Royal Society in the early years of its existence? Were there Cartesians and Gassendists in the early Royal Society?

Certainly, the sectarian division within the Society was commonly alluded to by Sprat's and Sorbière's contemporaries. Joseph Glanvill openly endorses the same judgment: there are Cartesians and atomists within the Royal Society, interested in the mechanics of collisions and the formulation of hypotheses.

Some of you (to whose excellent works the leaned world is duly indebted) publicly own the Cartesian, and the Atomical hypotheses.
(Glanvill 1665: Epistle dedicatory)

Unlike Sorbière, who was a Gassendist, and unlike Sprat, an advocate of the Baconian method, Glanvill himself seems to favor Descartes, at least in terms of methodology, over the atomists who are seen as making too-bold claims about the nature and constituents of reality.[11]

Thomas Sprat himself presents one of the leading virtuosi, Christopher Wren, as a mathematician and natural philosopher whose major work was the establishment of a *"Doctrine of Motion*, which is the most considerable of all others, for the establishing the first *Principles of Philosophy*, by *Geometrical Demonstrations"* (Sprat 1667: 311). In doing so, Sprat claims, Wren was following the footsteps of Descartes, in making collisions the basis for the whole of natural philosophy. Arguably, Wren was correcting Descartes's errors and works with the help of experiments and according to the new scientific methodology. However, there are some points of interest in Sprat's presentation of Wren and other mathematicians interested in the phenomena of collisions: they are seen as producing and testing hypotheses, working out the laws of collision, constructing a "doctrine of motion" and working toward the establishment of the principles of philosophy on geometrical demonstrations. They are, therefore, doing something rather different from the general program of gathering natural histories.[12] Furthermore, in establishing the laws of collisions as the basic elements of a mathematical physics, Wren is said to have established elements of "universal use": collisions being the basic "facts" of nature:

> Nor can it seem strange, that these *Elements* would be of such Universal use; if we consider that *Generation, Corruption, Alteration*, and all the Vicissitudes of Nature, are nothing else but the effects arising from the meeting of little Bodies, of differing Figures, Magnitudes, and Velocities.
> (Sprat 1667: 312)

Some of Boyle's writings from the same period also address the subject of Cartesians versus Gassendists within the Royal Society or inside the wider camp of the natural philosophy (Boyle 1999–2000, 6: 189ff). Boyle's attempt in the *Origin of Forms and Qualities* seems to be to get over the divisions between the two camps and consider them together under the name "corpuscular philosophers" while eliminating from natural philosophy the fundamental ontological assumptions concerning physical bodies (Jalobeanu 2006a).

Is there any substance to such a picture? Who are the Cartesians, who are the Gassendists? Are there really two groups, or even two camps within the Royal Society, maybe with different ideas about how science should be done? Are there Cartesians inside of the Royal Society?

We do have a serious problem here, concerning the labels: what does it mean to be a Cartesian? I will try to give a restricted answer in what follows, in the particular context of a well-known debate that took place within the Royal Society in 1668–1670 over the nature, laws, and proper treatment of collisions. It is a debate in which the sides are clear and the assent to a Cartesian program for natural philosophy easy to recognize. It is a debate that has Cartesians and Gassendists and even Cartesians turning Gassendists. And, despite being the subject of a considerable attention, it still has its less-explored characters and episodes.

As has been shown, the debate over collisions was a challenge to the "Baconianism" of the early Royal Society, posing methodological, epistemological, and metaphysical problems.[13] My purpose here is to explore some of these problems and to show that they have a common core: a general concern over the nature of physical bodies, and more specific methodological and epistemological concerns over the proper grounds for adopting hypotheses, adhering to a system of philosophy (whether Cartesian or atomist), and the relation between experiments, facts, and theories. I will focus upon a lesser-known character of the debate, showing what I claim to be a good example of the methodological, epistemological, and metaphysical issues emerging from the debate over being a Cartesian or a Gassendist in the Royal Society in the 1660s.

THE DEBATE OVER COLLISIONS AND THE LAWS OF MOTION

The debate over collisions and the laws of motion is usually presented as a debate between mathematicians or, at least, between those interested in mechanics: Christopher Wren, John Wallis, and Christiaan Huygens, with several side figures of lesser importance at the beginning, and then with other leading figures like Leibniz and even Newton joining in over the development of dynamics.[14] It is usually presented as the beginning of a longer and deeper controversy over *vis viva*, or the conservation of energy and the problem of forces in collisions. According to this story, the debate is interesting because it led to the realization of the fact that mechanics is incomplete without dynamical notions. It started around experiments with colliding bodies and continued with various small tracts where the laws of motion were worked out, all of them incorrect, of course, since none of the participants had the dynamical concepts of mass, inertia, or force.

What I want to do in the following sections is to show that the debate over collisions is interesting for reasons connected with bigger questions of the seventeenth-century natural philosophy concerning the proper methodology of science and the ultimate nature of physical reality. In many ways, the debate over the laws of motion in 1668–1670 is a debate over the nature of physical bodies within the "new" mechanical philosophy, with serious metaphysical, epistemological, and methodological consequences for the subsequent development of natural philosophy.

There are at least three different stages of the debate over collisions. The first stage involves various experiments over colliding bodies that seem to have sent three of the virtuosi to work on correcting Descartes's laws of motion. The second stage begins in 1668 with a series of queries formulated within the Royal Society and sent over to Wren, Wallis, and Huygens, in order to complete the mathematical model of collisions with a physical hypothesis concerning bodies. This is also the stage of the debate properly speaking, with new characters entering into the field of mechanics and with the formulation of new hypotheses on motion. There are two camps, broadly speaking: the mathematicians who seemed to have believed that the problem of collisions is solved once the laws of motion are formulated and tested; and the philosophers, who argued that a physical explanation of the phenomena is needed, together with improved definitions of the concepts employed to make sense of the behavior of two colliding bodies. The third stage of the debate is even more interesting because we can see some of the people involved in the debate realizing the extent and depth of the problems encountered. However, before telling the story of such a debate, it is important to understand its starting point: in what way are Wallis, Wren, and Huygens "Cartesians"?

DESCARTES'S PROGRAM OF REDUCING PHYSICS TO GEOMETRY

In the second part of the *Principles of Philosophy* Descartes attempts to reconstruct a theory of interacting bodies starting only from extension and its geometrical properties, namely from "a certain substance extended in length, breadth, and depth, and possessing all those properties which we clearly conceive to be appropriate to extended things" (Descartes 1991: 39). Those geometrical objects are the elements of a new physics that, besides explaining other things, also deals with punctual natural phenomena such as collisions. In Descartes's theory, collisions are the only real events in the physical universe.[15] All other phenomena, from physics, physiology, or psychology, are to be reduced to their basic constituents, collisions between parts of matter. In the second part of the *Principles* Descartes not only formulates the laws of nature but also claims to derive from them his famous rules of collision, which, supposedly, describe the behavior of individual bodies during encounters. However, the objects that enter in such interactions are geometrical shapes in motion, since Descartes claims to equate matter with extension.[16] By claiming that "there is absolutely nothing to investigate about this substance except those divisions, shapes, and movements" (Descartes 1991: 77) *and* that he can explain, with the sole help of geometry, all natural phenomena, Descartes had set the foundations of a longstanding problem of natural philosophy. If the world is made of geometrical shapes in motion, how can we make sense of apparent body–body interactions? Where are the (active) powers that bring about changes in the natural world?[17] And, more specifically, in what way can forces be ascribed to various bodies? Descartes speaks about forces of motion and rest, about bodies being stronger and weaker in an interaction,[18] about resistance and *determination*, all qualities that, apparently, cannot be deduced from his own conception of physical bodies as shapes in motion. Or, if they can, the reconstruction involved is rather complex and implies a new mathematical approach.[19] In order to have a coherent account of reducing physics to geometry, Descartes is forced to claim that there is a way in which we can make sense of forces through pure mathematics.[20] I will call his claim "the strong programme for a mathematical physics."[21]

It has been repeatedly shown that Descartes's programme was a failure. It failed to give a proper definition of a physical body, which, in Descartes's view, is no more than parts of matter moving together (at rest with respect to each other).[22] It failed to give a definition of motion, which is considered to be both relative, and as belonging to a body.[23] It failed to explain the transference of motion from one body to another during collisions.[24] It also failed to make the connection between the laws of nature and the behavior of individual bodies in the encounters and it formulates highly counter-intuitive rules of collisions.

Although recognized as a failure, Descartes's program for a mathematical physics was highly influential. It appealed to mathematicians and philosophers alike because of its elegance and simplicity and because it postulates a simple and intelligible universe, a passive matter and a continuous interaction between God and His creation. It is in this sense that all the mathematicians working on collisions in the 1660s are continuing Descartes's program for a mathematical physics. They claim the possibility of making sense of physical interactions with the sole help of geometry and quantities, without being forced to reintroduce forces, occult qualities, or unintelligible properties like substantial forms, sympathies, active principles, and so on.

However, it is the very definition of a physical body and its qualities that is problematic within Descartes's program. A shape in motion has measure (of its surface), direction of motion, and speed (difficult to define in an instant, however, but, again, capable of being represented through a mathematical quantity). It does not have density, or mass, or inertia, or any dynamical quality at all. Indeed, in Descartes's view, the very definition of motion is rather weird: a body is said to be in motion if its surface is in a process of separation from the contiguous bodies and it is at rest otherwise.[25] Rest and motion are opposed states (*Principles* II, art. 44; Descartes 1991: 63) and there seems to be no way to communicate motion from one body to another. The way out of this paradox seems to be the postulation of many motions within each body. However, the various motions and tendencies that are opposed to each other (motion, rest, slowness, and swiftness, or opposed directions) cannot coexist in the same place. According to Descartes, when such a situation occurs, there happens a "change of modes" according to a supplementary requirement of economy or an attempt to "minimize the total change" (*To Clerselier*, February 17, 1645, AT IV 185).

The treatment of collisions, therefore, is reduced to a very peculiar process of appearance and disappearance of new bodies. Each body is a surface with certain associated instantaneous measures, or numbers, while motions or collisions are described in terms of the coincidence or separation of surfaces. If a body A meets a body B, what actually happens is a tendency of the two surfaces to become one, when in the new body so formed some contrary tendencies are fighting each other. The mathematician needs to compare them and decide what would be the rational outcome of such a collision, namely, if the new bodies appearing after the collisions are formed through a process of "reflection" or through a process of "absorption."[26]

When obliged to talk about forces, though, Descartes speaks in terms of tendencies or endeavors to preserve the state in which a body is (namely rest or motion), relating to the three laws of motion which are, in turn, grounded in God's immutability. The force of motion or the force of rest are no more than such numbers ascribed to the surfaces in motion, instantaneous and in many respects conventional: they are the means through which the mathematician has the possibility of keeping track of a body through its "measures" and by assigning positions to it in a Cartesian space (another mathematical fiction) while comparing it with other bodies again in terms of the measures (number) involved.

As Huygens himself had stressed, there is something deeply compelling in Descartes's program, even if, in itself, it fails to make sense of the actual motions, or the actual phenomena of the universe.[27] It allows mathematicians to work on mechanics and, providing they find good definitions for the qualities involved in the collisions (like hardness, elasticity, and bulk, for example), there is a way to construct a well-founded theory, based on reason. As a program for a mathematical physics, Descartes's natural philosophy is compelling and interesting, at least according to Huygens, because it proposes the first complete "mechanical system."[28] On the other hand, this mechanical system offers only a tentative, incomplete, and unfinished path for further investigation of the physical world. In claiming the truth of his principles, Descartes has damaged the whole enterprise, founding a "sect" instead of a theory (Huygens 1888–1950, 10: 405).

It is interesting that Huygens gives us a "good meaning" and a "bad meaning" of what it is to be Cartesian. It is good to be a Cartesian if you follow Descartes's attempts to establish a mathematical physics but keep in mind the tentative and fallible character of the whole enterprise. According to such a reading, one can start along Cartesian lines from extended matter and geometry, and continue to develop a mathematical physics without repeating Descartes's mistakes; most importantly, without mistakenly taking one's own hypotheses as truths about the physical world. This is exactly what Oldenburg or Sprat mean when claiming, for example, that Wren or Wallis are Cartesians.

In this sense, then, there were a good number of Cartesians in the Royal Society, working along the same lines of Descartes's project of reducing physics to geometry, correcting the laws of collisions and trying to find the laws of matter in motion. Wallis, Wren, and Huygens embarked upon this project, starting with a similar mathematical model of bodies in motion. For all three, bodies are geometrical objects moving in the geometrical space, endowed with a minimum of physical qualities that they hoped to reduce to a mathematical description. Such qualities were the bulk, density, hardness, and the elasticity of bodies. However, if the first two qualities were measurable, through experiments (weighing the body to determine its "bulk," and measuring relative densities) the last two were mathematical abstractions. "Hardness" could mean a property of the way in which particles are packed inside the body, or, simply, the property of suffering no distortion upon impact (Wallis). Therefore, the controversy between Wallis, Wren, and Hooke, for example, started over the definition of "hardness" and the correspondence between the mechanical definition and the physical reality inside bodies.

EXPERIMENTING WITH COLLISIONS AND THE PROBLEM OF FINDING THE CORRECT LAWS OF MOTION

The problem of collisions and experimenting with colliding bodies is listed among the Royal Society's achievements in Sprat's *History*. In Sprat's view, the initiator of a doctrine of motion within the Royal Society was Christopher Wren. He was the one who corrected Descartes's errors, putting the foundation for a new field of study. The resulting "Doctrine of Motion" was established upon many experiments and the invention (by Wallis) of a special instrument designed for the trial of various collisions (Sprat 1667: 312). It provided the virtuosi with a general theory about the basic elements of nature, since collisions are seen to be the very origin of all the phenomena in a mechanically constituted universe.

The role of the Royal Society in the exploration for the "true laws of matter and motion" features equally prominently in the works of the other apologist, Joseph Glanvill. In Glanvill's introduction to *Scepsis Scientifica*, for example, finding the rules of collisions, i.e. the "true laws of motion," becomes one of the main purposes of the experimental activities of the Royal Society (Glanvill 1665).

There seems to be a twofold purpose of the Royal Society in Glanvill's and Sprat's visions: one is the gathering of facts, the other is the construction of theories, based upon geometrical demonstrations and experimental confirmations. The problem of collisions is of the latter kind. Meanwhile, the importance of the collisions and laws of motion is tremendous. By showing that:

1. everything in the universe can be reduced to collisions between material particles
2. the very investigation into the causes of the material world means no more than finding the "rules of Actions and Passion among the parcels of the Universal Matter" (Boyle 1999–2000, 3: 245)

the virtuosi were able to construct a theory of motion within the general framework of the mechanical philosophy.

However, which mechanical philosophy? There were a number of mechanical philosophers in the seventeenth century disputing over fundamental definitions of space, time, matter, motion, bodies, hardness, elasticity, or force.[29] This is exactly what was at stake in the debate over collisions. If the first phase of it concerned the early experiments and some work done by Wren and Wallis, as presented in Sprat's history, the second phase, or the debate properly speaking, began in late 1668, when at a meeting of the Royal Society, Hooke seems to have raised the general problem of collisions as one of the unsolved problems to be pursued by the virtuosi.[30] He seems to have encountered difficulties in setting the agenda of the Royal Society over the problem of collisions. This is at least how Oldenburg presents the problem when writing to Huygens and Wren, and asking them to send their theories about motion and collisions. In the letter to Wren of October 29, 1668, Oldenburg tells the story in the following way:

> On Thursday last at the publick meeting of the R. Society it was proposed by some, that there might be made experiments to discover the nature & laws of motion, as the foundation of Philosophie and all Philosophical discourse, to which proposall when it was mentioned by others, that both you and Monsr. Huygens had considered that subject more than many others, & probably found out a Theorie to explicate all sorts of experiments to be made of that nature, I was commanded to desire you, as well as Monsr. Huygens, in the name of the Society, that you would pleas to impart unto them what you had meditated & tryed on the said argument, assuring yourself, that these communications of yours shall be registered by the Society as your productions, and stand in their booke as one of the best monuments of your Philosophicall Genius.
> (Oldenburg 1965–1986, 5: 117–118)[31]

Not only was Oldenburg asked to write to Huygens and Wren, but at the following meetings of the Royal Society, in November 1668, the problem of collisions seems to have gained unprecedented support, and to have developed into a sort of research program.

> It was also moved, that since the Society was upon the disquisition of the nature, principles, and laws of motion, all authors, who had written on that subject, and delivered their hypotheses concerning it, might be consulted and examined, and an account of their opinions brought in, to see, what had already been done in this matter. Whereupon Mr. Collins was desired to peruse such authors, and particularly Des Cartes, Borelli, and Marcus Marci: and Mr. Oldenburg was desired to write to Dr. Wallis, that he should take a share of this work.
>
> (Birch 1756–1757, 2: 320)

Note, again, that the structure of the research program presented here is specifically against the advocated methodology of the Royal Society: searching in books, not in nature, gathering hypotheses, not experiments, pursuing further the hypotheses and testing them with new experiments.[32] This is also how Oldenburg presents the problem in his letter to Huygens, whom he asks to send not his experiments and observations but his doctrine and laws of motion to be tested and confirmed by the virtuosi (Oldenburg 1965–1986, 5: 103).

Huygens, Wren, and Wallis eventually sent their respective works to the Royal Society to be recorded, discussed, tested, and published. The three theories concerning collisions were read before the Society[33] in November–December 1668.[34] However, the most important part of the debate was just beginning. Although Oldenburg claimed, in numerous letters, that Huygens's and Wren's laws of motion had been shown to be similar, and, therefore, that there is no debate,[35] the problem of collisions remained central to his correspondence for at least the next two years. And the next stage in the story is the beginning of the controversy itself. It is not a controversy about the correct laws of motion, but about something more general, touching the basic concepts and methods of physics. It began with the formulation of a set of queries concerning collisions that were forwarded to Wallis, Wren, and Huygens. Then, new characters entered into the debate, either claiming that the Royal Society does not have a proper theory of motion, despite the formulated laws, or that something even deeper and more troublesome is at stake here: the proper methodology for doing science, the basic concepts or principles of natural philosophy, or the fundamental ontology. Why is this so? Let us have a look at the set of queries formulated as the outcome of the first two steps in the debate.

Queries Concerning Bodies and Motion

On December 1, 1668, Oldenburg wrote to Wallis, sending him a list of questions, established at the previous meeting of the Royal Society (a week before). The list of queries contains rather general questions concerning the nature of motion and rest, and, most important, the nature of physical bodies. The first two deal with the definition of hardness, elasticity, and resistance to motion. The next three address questions like the transfer of motion from one body to another, the conservation of motion, and the possibility of destroying motions through collisions. In presenting the debate, Oldenburg stressed the fact that:

> the Society in their present disquisitions have rather an Eye to the Physical causes of Motion, & the Principles thereof, than the Mathematical Rules of it.
> (Oldenburg 1965–1986, 5: 221)

Now, this is something new, since it makes a strong separation between the mathematical treatment of motion and the search for physical causes. The queries were sent to the mathematicians, asking them to complete their theories with a physics of motion.

There were five main points in question:

1. Whether springiness is the only cause of rebounding (which means that a definition of hardness and elasticity is looked for in terms of causes, therefore, appealing to the composition and structure of bodies)
2. Whether "quiescent Matter" has any resistance to motion? This is a question to which Descartes responded in an affirmative way, while all others to this point had responded negatively, as Wallis, for example, points out. (Wallis to Oldenburg, December 1, 1668, Oldenburg 1965–1986, 5: 220–222). The problem of force of resistance is simply that something else over and above matter and motion is required to explain how a body at rest can have a force at all.[36]
3. Whether motion may pass out of one subject into another[37]
4. Whether no motion in the world perish, nor new motion be generated[38]
5. Whether different motions, meeting, destroy one another (a corollary of the previous query, leading again to metaphysical questions).

There is something really peculiar in the formulation of such queries about motion. They seem to start from a concern about the proper definitions of things like hardness and elasticity. However, they reintroduce into the debate metaphysical questions, a point not lost on Huygens, Wallis, and Wren. The questions regarding the creation and annihilation of motion belong to well-known and knotted issues of metaphysics and natural theology. The queries also touch on the relation between mathematics and natural philosophy[39] and are dangerously close to problematic questions like the nature of the fundamental constituents of reality. They are all that the Royal Society fears and apparently rejects. However, *they* remain the focus of the debate for at least two more years. Oldenburg sent the queries around and collected answers from various parties, while also looking for possible alternatives to those in hand.[40] And at this point two new characters entered the debate. One, better known but less conspicuous in Oldenburg's correspondence, is Robert Hooke, who might have been the instigator of the whole debate. We know that Hooke was preoccupied with finding the causes of elasticity and hardness[41] and had a different definition of hard bodies from the ones suggested by Wallis or Wren. Hooke famously asserted that in a collision between two hard bodies, both will come to rest.

The second character, less well known today, is William Neile. Neile's name comes up frequently in Oldenburg's correspondence of 1669. He was the author of a hypothesis on motion discussed extensively by Wallis, he seems to have been behind the queries themselves, and he also seems to have been a sort of driving force behind the whole issue of constructing a physics of collisions. In fact, the amount of attention Neile's hypothesis received from Oldenburg and Wallis is peculiar. Oldenburg sent it around to Wren, Wallis, and Huygens and pursued the matter further, until even Boyle himself became involved in the debate. Wallis, who refused to discuss the problem of collisions with Huygens, and was not always very friendly (or even civil) with his other correspondents, wrote at least five extended letters discussing and carefully refuting Neile's hypothesis on motion. So, who was William Neile and why did he have such a prominent role in the debate?

WILLIAM NEILE'S HYPOTHESIS ON MOTION AND HIS ROLE IN THE DEBATE

Little is known about William Neile (1637–1670), son of Sir Paul Neile, one of the founding members of the Royal Society (Ronan and Hartley 1960) and grandson of Richard Neile Archbishop of York. He had entered the Wadham College in Oxford in 1652, presumably working under the supervision of Wilkins and Ward, since we know that his mathematical talent was precocious. He matriculated in 1655 and by 1657 he was at Middle Temple, in London, studying law. In the same year, at the age of nineteen, he gave an exact rectification of the cubical parabola, a discovery communicated to Wren, Brouncker, and the others at Gresham College.[42] In 1662/3 Neile was elected as a Fellow of the Royal Society and in 1666 joined the Council. In 1669 he developed a theory of motion in a series of successive letters to Oldenburg and his hypothesis is registered in April 29, 1669, in the register of the Royal Society (Ronan, Hartley 1960). There was a considerable debate over Neile's hypothesis throughout 1669 and early 1670. Neile also seems to have been making astronomical observations and he was a member of King Charles's II Privy Council. However, his career ended abruptly on August 24, 1670, when, according to some stories, he died rather mysteriously. This is how the English historian Thomas Hearne describes William Neile:

> He was a virtuous, sober, pious man and had such a powerful genius to mathematical learning that had he not been cut out off in the prime of his years, in all probability he would have equaled, if not excelled, the celebrated men of that profession. Deep melancholy hastened his end, through his love for a maid of honour, to mary whom he could not obtain his father's consent.
> (Hearne 1764, 5: 142)

A brilliant young mathematician, a powerful family, a romantic love story and an early death; enough elements for a vivid portrait; all these, however, do not explain Neile's prominent position within the debate over collisions in the year 1669. Was it the hypothesis itself that was important? I would like to show in this section that what was really important here was not Neile's hypothesis but his persistent pursuit of a physics of collisions and his insistence upon continuing the debate long after many of the virtuosi considered the problem solved.

In his letters to Oldenburg, Neile presented himself as a mathematician and he claimed he had no desire to become a philosopher. For example, in a letter from January 21, 1668/9, he claims:

> for my owne part if the principles I offer should prove never so much true yet I could not wish for the name of a Philosopher if I could have it it is so troublesome a name that and the name of Mathematician is almost as dangerous as the name of a Poet (which Ben Jonson mentions).
> (Oldenburg 1965–1986, 5: 344)

Neile's reference points to the traditional Aristotelian meaning of *poien*: to make or to feign objects.[43] Similarly, Neile claims, the Mathematician is feigning the objects of his study. What then does a philosopher do? I think that what we have in this letter is an interesting glimpse into what looks more and more like a dramatic understanding of the depth and extent of the problems involved in the debate over collisions.

However this may be, Neile is not only interested in the physics of collisions but obviously deeply unhappy with the solutions offered by Wren, Wallis, and Hooke. He began to correspond with Oldenburg and, in a series of short letters from December 1668, he continuously pointed out that the problem of collisions was unsolved because no one was able to formulate a general hypothesis to explain the phenomena and ground the definitions of hardness and elasticity. He asked for Hooke's hypothesis, Wallis's, Wren's, and Huygens's theories, and he was also, most probably, one of the originators of the queries on motion.

Neile's position is extremely interesting. In every single letter he begins by saying that we do not know the definition of the most basic concepts involved in demonstrating the laws of motion and that the whole phenomenon of collisions is deeply problematic (December 18, 1668, Oldenburg 1965–1986, 5: 263–264). Then he formulates a hypothesis concerning the nature of bodies, in reply to the same queries, stating that bodies are composed of particles in an innumerable number of motions, in all possible directions. Resistance and hardness, according to Neile, are not the result of rest between particles, but must result from particular kinds of motions.

> A great disadvantage. for Mr. Hooks opinion that if a body were perfectly hard it would have no spring at all I think a body cant be made hard without motion in its particles that is with out a spring and the more motion it has the more spring it has, that is the harder it is the more spring it has and for my part I think that a diamond (or what ever body is the hardest in nature) has a stronger spring then other bodies and a greater quantity of motion in it. I think all bodies are like fire only a masse of particles variously moving and sometimes resting alternately.
> (Oldenburg 1965–1986, 5: 264)

However, when Oldenburg insistently asked him to send his hypothesis on motion, Neile seemed to be rather reticent at first, claiming that he was no philosopher and that others, more qualified, should be in charge of drawing up such a hypothesis. He claimed, though, that we cannot elaborate a theory of motion and collision without a hypothesis about the nature of bodies and a good definition of hardness and elasticity. Moreover, such a theory has to start from answering the two important queries: the one concerning the resistance of a body at rest, and the one concerning the conservation of motion, the two most difficult questions of the whole set of queries. In Neile's opinion the whole debate raises deeper problems. The formulated laws of motion are just the tip of the iceberg:

> Dr. Wren I think he assumes his axiome a great deale sooner then he need to doe for if it be possible that the nature of motion be really no otherwise then it seemes to be by experiment to conclude that the aparence is the reality and that the aparence must not be denied to be really true under pretence that it is an axiome meethinks is not very philosophicall.
> (Late January 1668/9, Oldenburg 1965–1986, 5: 363)

Neile seems to have believed not only that we need a definition of body and its properties in terms of a matter theory in order to be able to speak about collisions, but also that the whole issue bears upon constructing a theory of knowledge:

> most people think that matter in motion has a repugnancye or inaptitude to be stopped I can't find any reason for it for the aparence of the things signifies nothing to the business aparence relates to the whole compound bodie but the motion of the simple particles is invisible.
> (January 16 and 22, 1668/9, Oldenburg 1965–1986, 5: 347)

What happens when two bodies collide is a peculiar motion of the small particles that compose the two bodies. What we can see when experimenting with collisions is not what really happens, but a mere appearance. This is strikingly similar to Descartes's conception of a mathematical physics in which what happens are appearances and disappearances of shapes, while what we see are bodies in motion. Neile claimed that experiments are not enough for the knowledge of nature, that what we need is a deductive science based upon general principles.

> I desire to know the nature of motion and the nature of quiet I desire not only to know that if there be two bodies of a considerable magnitude moving against one another they shall reflect with such a swiftnesse for that they may doe and yet motion may not reflect from motion when it moves with it and to know a thing barely by experiment is good for use but it is not science or philosophye I can say no more but that a good cause may many times suffer for the badnesse of the defender or because people are not willing to trouble themselves to consider it or to consider it with prejudice.
> (Neile to Oldenburg, May 7, 1669, Oldenburg 1965–1986, 5: 518)[44]

Therefore, explaining collisions meant for Neile:

1. finding a model or a good definition of physical body and its qualities based upon a theory of matter in motion
2. discussing the collisions not in terms of appearances, but in terms of the real motions of particles inside the bodies

Moreover, although both hypotheses and experiments are necessary, the strongest condition for a good theory is the intelligibility of the hypothesis:

> so unlesse I could as cleerly describe the motion of small particles as if I had seen them through a chrystall glasse moving like so many flyes all I can say is but obscurity.
> (Oldenburg 1965–1986, 5: 517)

Neile's hypothesis on motion was based, then, on imagining bodies as streams of particles in perpetual motion. The observed phenomenon, the apparent resistance of a body at rest, arises, then, from a coherent motion of the component particles:

> Since it may be gathered from what has gone before that wherever there is resistance, or a body's reaction, there motion is to be found, for that is the unique cause able to produce such an effect naturally because of the incapacity of matter at rest to impede motion in any way . . . and since it is very obvious that all bodies known to us possess this faculty of resisting more or less, it is on that account quite fitting that we attribute motion to them.
> (Oldenburg 1965–1986, 5: 525)

In each body there is an almost infinite variety of motions, which, for a body at rest, happens in any direction.[45] However, once a body is in motion, the internal motions of particles form a flux in the direction of motion.[46] Meanwhile, all the particles preserve other motions, too, in some other directions. Neile seems to have believed that the direction of motion and the motion itself vary independently, much like Descartes's motion and *determination*. So, in each body, there are innumerable motions on various "lines of motion" (*plaga*). Moreover, each motion may be interrupted by minute periods of rest.[47]

How can we explain collisions starting from this hypothesis? Neile begins again from a Cartesian point of view: the standard cases of the collisions being reflections or absorption of the incoming stream of particles. Each collision happens along a line of motion and only those microscopic motions that take place along the line are taken into consideration. For the frontal collision of two bodies, one in motion, the other at rest, what really happens is that each particle at rest is put into motion and that each two colliding motions annihilate each other. Again, what counts is the direction of motion: the frontal collision affects only the motions taking place in that direction of motion. Meanwhile, the motions of the small particles along other directions of motion remain and, after the collisions, take control of the whole body. This is how reflection is described:

> For as that stream of particles (so to speak) is suddenly obstructed in its former course it must bounce back with an equal speed, since that sudden loss of motion towards the left while at the same time the propulsion towards the right continues (because it is not at all impeded), the same preponderance of motion towards the right remains, as there was formerly towards the left. Thereby the body S will be carried back towards the right with the same velocity with which it first approached to the left, and for the same reason the body R will in a similar way return towards the left.
> (Oldenburg 1965–1986, 5: 526–527)

Neile illustrates his hypothesis with an interesting analogy: we can view the whole situation as a battlefield on which there is a war between particles: each particle colliding with another along the same line of motion is "killed" and carried along. If the collision is frontal, both particles stop and are carried along by the surrounding particles. As a result, there is a continuous destruction of motion in the world.[48]

Neile's hypothesis is developed in the entire series of letters, starting from April 1669 and elaborated further under the constant pressure of Wallis's refutations. Wallis's side of the correspondence is also extremely interesting; first, because the letters refuting Neile's hypothesis are the longest letters of Wallis from that period. Even more amazing, however, is Wallis's patience in explaining again and again every detail of his refutation, which is, in fact, based upon the fact that in Neile's vision, motion is constantly decreasing, no new motions are created and this will eventually annual all the microscopic motions. There are at least three long letters from Wallis to Oldenburg repeating his main argument, with more and more details. Moreover, Wallis does not seem to be completely opposed to the formulation of such a hypothesis. There are a number of points on which he agrees with Neile: starting from the need to have such a hypothesis in terms of real particles and real motions. Wallis agrees that:

1. the missing definition of body and its qualities is a problem for natural philosophy and for the attempt to find laws of motion
2. we do not know the causes of motion or the force of motion (*vis motrix*) and we have no explanation for resistance
3. such an explanation must begin with postulating real motions, not merely apparent or relative motions.

At this point, I suggest, the debate moved beyond its Cartesian beginnings, with both Neile and Wallis changing camps. There are a number of arguments for such an interpretation and I will just sketch them here.

First, Huygens, who was still working along the same Cartesian lines, formulated only one comment in reply to Neile's hypothesis: that it is too metaphysical and that it is wrong in talking about the "real motion" of the particles. In reply, Neile introduces a famous example: what happens in the case of a body that is alone in the universe?

This question is taken further by Oldenburg himself, whose place in this debate has not been studied so far and who, I think, is equally interested in finding an answer as the others. In his letters, he sent around a reference: Boyle's treatise on absolute rest.

And this is my second argument: Boyle's treatise on absolute rest is an appendix to the second edition of *Certain Physiological Essays*, published in 1669. It has an interesting "advertisement" to the reader, in which Boyle claims that the tract was born from a discussion with other virtuosi, who were interested in the problem of bodies, motion, and rest and who asked him to write something on the subject, to correct the errors of his previous tracts on fluidity and firmness (Boyle 1999–2000, 6: 191). Such a conversation can be seen as relating most certainly to the debate over collisions. Might it not have been Oldenburg himself the virtuoso who had asked Boyle to take a stand in the debate?

The third argument is really an example to round off this long story. In recognizing his failure to account for the nature of bodies, Neile turned to the necessity of a definition of absolute motion. Then, invited to give such a definition, he actually closed his correspondence on the subject with a quotation. Neile, the mathematician who did not want to be a philosopher, but turned Cartesian out of necessity, closed his letter quoting from Gassendi on the subject of forces, individual atoms created in the void space, and their motions within the physical universe.

CONCLUSION

It was my purpose to show in this chapter that the debate over collisions within the Royal Society started from a Cartesian program for a mathematical physics. It then gradually evolved into a different kind of debate, with metaphysical and methodological overtones. On one hand, the queries on collisions opened the discussion concerning the missing definition of bodies and the mechanical reconstruction of qualities, something that was definitely at the borderline of the famous divisions between the Cartesians and the Gassendists. On the other, the debate addressed important questions concerning the role of experiment and hypotheses. It is not an accident, I think, that Neile's next (and last) paper, was a proposal for the reformation of the Royal Society. In it, a new methodology was sketched, around the searching for causes and testing hypotheses. A committee was proposed, for example, to search for the causes of the experimented phenomena. The true science, in Neile's view, had to be a science of causes, while experiments served only as possible tests for it.

I also wanted to show that the debate over collisions engaged more people than are usually discussed: and that at least Oldenburg and maybe Boyle were also interested in it less because of its mathematical aspects and more because it related to the general "scandal" of the mechanical philosophy in the 1660s: the "missing definition of bodies."

NOTES

1. For example, Michael Hunter identifies at least two Baconian strands within the Royal Society. One is Bacon's project of the *Sylva Sylvarum*: a random collection of facts, observations and experiments, open to anyone, a sort of catalogue for future natural histories. Another one is the Baconianism of *Novum Organum*, a sort of eliminatory process of induction, where formulating hypotheses, or anticipations, and testing them is an important part

of the method. Michael Hunter 1989: 207–209. See also M. B. Hall 1991; Hunter 2007.
2. Sprat's *History of the Royal Society* abounds in paragraphs emphasizing the collective, tentative, and introductory character of the experimental enterprise: the virtuosi are not required to be "perfect philosophers," not even philosophers *per se*, but "plain, diligent and laborious observers" (Sprat 1667: 72); "sincere witnesses standing by" (Sprat 1667: 73). See also Sprat 1667: 85.
3. According to the statutes quoted by Sprat, the Royal Society has "to order, take account, consider, and discourse of Philosophical Experiments, and Observations: to read, hear, and discourse upon Letters, Reports, and other Papers, containing Philosophical matters, as also to view, and discourse upon the productions and rarities of Nature, and Art: and to consider what to deduce from them, or how they may be for use, improv'd, or discovery," Sprat 1667: 145. In Sprat's book the Royal Society is presented as being explicitly concerned to avoid both dogmatic philosophy, scepticism, and sectarianism, Sprat 1667: 101. "They have attempted, to free it from the Artifice, and Humors, and Passions of Sects; to render it an Instrument, whereby Mankind may obtain a Dominion over *Things*, and not onely over one anothers *Judgements*," Sprat 1667: 62. See also Sprat 1667: 64–66 for the dangers of the philosophical schools, for the ancient and for the modern philosophy alike.
4. "'till there be a sufficient Collection made of Experiments, Histories and Observations, there are no Debates to be held at the Weekly Meetings of the Society concerning any Hypothesis or Principle of Philosophy, nor any Discourse made for Explicating any Phenomena, except by special appointment of the Society or allowance of the President," Hunter and Wood 1986: 66.
5. One thorough description of this connection can be found in Sprat's description of the way in which the adoption of a general hypothesis or theory inflames the imagination and leads to a distorted investigation of nature. The experimenter who starts with a hypothesis or a causal theory cannot be a faithful experimentalist anymore: "he meets with more and more proofs to confirm his *judgment*: thus he grows by little and little, warmer in his *imaginations*: the delight of his success swells him: he triumphs and applauds himself, for having found out some *important Truth*: but now his Trial begins to slacken: now *impatience* and *security* creeps upon him: now he carelessly admits whole crowds of Testimonies, that seem any way to confirm that *Opinion*, which he had before establish'd: now he stops his survey, which ought to have gone forward to many more *particulars*; and so at last, this *sincere*, this *invincible Observer*, out of weariness, or presumption, becomes the most negligent in the later part of his work, in which he ought to have been more exact," Sprat 1667: 103. For a more general discussion of the experimental philosophy in seventeenth century see Anstey 2005.
6. Jalobeanu 2008.
7. Sprat 1667 part. II sect. XVII–XVIII.
8. The book was translated into English and went through a number of editions, usually bounded together with Sprat's and Wren's answers to Sorbière's critiques of the Royal Society. See Sorbière 1709; Sprat 1708.
9. The whole episode is very interesting and very illuminating for the early history of the Royal Society. To the modern eye, Sorbière's account of England is vivid and only moderately ironic, while the description of the Royal Society is sometimes enthusiastic. True, he has a number of colorful portraits of the virtuosi; however it is not entirely clear why the book was perceived as Voltaire described it "a dull scurrilous satire upon a nation he knew nothing

of" (quoted in Syfret 1950: 54). As for Sorbière's account of the meetings and fellows of the Royal Society, they might be seen as interesting and useful, although with the reservation that their author did not speak English. Sprat's answer, on the other hand, seems entirely disproportionate and, concerning the Royal Society, not up to the point. I would like to suggest that, maybe, by presenting an image of the virtuosi discussing theories and debating Cartesianism versus Gassendism, and by showing the picture of a Society divided between intellectual parties, Sorbière was seen as dangerous for the image of the Royal Society (that Sprat was willing to put forward in his book). See also Lennon 1993: 101–103 for a similar view on Sorbière's book and Sprat's maltreatment of it.

10. "He first says that they are not all guided by the Authority of Gassendus and Descartes; but that the Mathematicians are for Descartes, and the men of General Learning for Gassendus; whereas neither of these Two Men bear any sway among them: they are never named there as Dictators over Men's Reason; nor is there any extraordinary reference to their Judgements," Sprat 1708: 241; Sorbière 1709: 165. For Gassendi's reputation as "literatus" in seventeenth-century England, see Joy 1987.

11. The case of Descartes's philosophy, the "only way to science" is defended in strong terms in Glanvill's first book, *The Vanity of Dogmatizing*, 1661. In the rewritten version of the book, dedicated to the Royal Society in 1665, Glanvill plays down his open advocacy of Descartes's method, Descartes's philosophical reasoning, and mechanical cosmology. However, most of the modifications concern the style and no major argument in favour of the Cartesianism is eliminated. See also Glanvill 1665.

12. Sprat is careful to distinguish the use of hypotheses within the Royal Society as being of a different kind from the hypothesis of the dogmatic philosophy. For example, when enumerating the published works of the virtuosi, many with the word "hypothesis" in their title, he states that: "In this Collection of their *Discourses*, and *Treatises*, my Reader beholding so many pass under the name of *Hypotheses*, may perhaps imagine that this consists not so well with their Method, and with the main purpose of their *Studies*, which I have often repeated to be chiefly bent upon the *Operative*, rather than the *Theoretical Philosophy*. But I hope he will be satisfied, if he shall remember, that I have already remov'd this doubt, by affirming, that whatever *Principles*, and *Speculations* they now raise from things, they do not rely upon them as the absolute end, but only use them as a means of farther *Knowledge*," Sprat 1667: 257.

13. A. R. Hall 1966; M. B. Hall 1991. For a more recent account of the way in which the debate has shaped seventeenth-century physics see Bertoloni Meli 2006.

14. See, for example, Fichant 1978, 1993; Merchant 1967.

15. Fichant 1993 emphasizes the importance of Descartes for the assumption that all physical phenomena can be reduced to collisions.

16. *Principles*, II. arts 4, 64; AT VIII 42, 78–79. See also Woolhouse 1994; Garber 1992.

17. Such questions have been the subject of much attention in the last ten years. See for example the debate between Des Chene 2000 and Menn 2000. For a survey of the problem see Hattab 2000. On the more general questions of body–body causation (and hence the problem of forces in Descartes's physics) see Pessin 2003; Schmaltz 2003; Freddoso 1991. On the laws of collisions and forces in Descartes's physics see Slowick 2002; Jalobeanu 2006b.

18. Alan Gabbey (1980) has labeled Descartes's account of collisions the "competing model": the first purpose when studying an encounter is to decide

which of the bodies is "stronger." Such a decision, however, involves a complicated process in which one cannot compare the force of action with the force of resistance, e.g. the force of the moving body with that of the body at rest and, moreover, Descartes seems to apply a principle of minimizing the change in the state of the body (see Descartes's famous letter to Clerselier, February 17, 1645, AT IV 185).
19. Jalobeanu 2006b.
20. *Principles*, II, art. 45; AT VIII 67.
21. See Gaukroger 1980b, for a definition and explanation of such a programme.
22. The correspondent definition of motion is given using the definition of "body," creating thus a circle; *Principles* II, art. 25; AT VIII 53–54.
23. See Gaukroger 1995 for a contextual interpretation of this contradiction: Descartes began with a conception of real motions and then replaced them with relative motions because he was afraid of the reactions of the Church. Garber (1992) suggests that we should rather look at the definition of motion as "relational" (and not relative). For an interesting reconstruction of Descartes's conception of motion see Des Chene 1996.
24. This is a very important subject in Descartes's correspondence with Henry More, who emphasizes that if motion is a mode of a body, than it cannot be said that it passes from one body to another. See *To More*, August 1649, AT V 404–405.
25. See *To Clerselier*, February 17, 1645, AT IV 187, for the most clear explanation of what happens at a collision and the subsequent definition of motion.
26. Jalobeanu 2006b. I am also following here the interesting rational reconstruction of Descartes's collisions proposed by Des Chene 1996.
27. There is an interesting assessment of Descartes's natural philosophy and its shortcomings in Huygens's memorandum on Baillet's biography of Descartes, where, after claiming that Descartes's natural philosophy is compelling because it is a sort of novel, Huygens insists upon the intelligibility and simplicity he finds compelling in Descartes's physics: "What was most pleasing in the beginning, when this philosophy appeared [in print], was that one could understand what M. Descartes was saying, unlike other philosophers who gave us words without signification, like 'qualities,' 'substantial forms,' 'intentional species,' etc" [my translation]. However, Huygens continues, Descartes replaced the authority of the ancients with the authority of his own causal hypotheses, Huygens 1888–1950, 10: 403.
28. "He should have proposed his system of physics as a provisional [explanation]: showing what can be tentatively achieved in this science if one only admits in it the principles of mechanics. He should have invited others to join in this research. Such an enterprise would have been highly commendable. But, in wanting to make us believe that he had discovered the truth . . . he created a thing which is of great prejudice for the progress of philosophy, because those who believed him and became his followers, have imagined they have the knowledge of all the causes that can be known and so they are often losing more time in following the doctrine of their master than in trying to discover the real reasons behind the numerous phenomena of nature on which all Descartes had to say were mere fictions," Huygens 1888–1950, 10: 405.
29. Descartes, Hobbes, Galileo, and Gassendi are, in a way, all mechanical philosophers. However, they constructed very different brands of mechanical philosophy, starting from different ontologies and with different definitions of the fundamental concepts. As a result, the mechanical philosophy is usually presented, by the middle of the seventeenth century, as a field in a deep crisis,

a battlefield of sectarian philosophers quarrelling over the nature of bodies, the ultimate constituents of reality like atoms, void, space, time, continuity versus discontinuity, etc. There are numerous passages in Glanvill's *Scepsis Scientifica*, 1665, which summarize the debate very vividly and the debate over the nature of bodies will clearly appear in what follows as being one central concern of those involved in the attempt to find the laws of motion.

30. Experimenting on collisions did happen from time to time within the Royal Society before this date. In May 1668 Oldenburg mentioned the "theory of mechanics" among the chief interests of the Royal Society, Oldenburg 1965–1986, 4: 424. In July 1668 Oldenburg wrote to Boyle that "Lately we fell upon the examination of pendulums which was occasioned by a Proposition advanced by Borelli, *De vi percussionis*, who seems to assert; That the line of a Pendulum being stopp'd by a pin or other thing in the Perpendicular line any where, the bullet holds on its motion beyond the perpendicular (though in another Circle) to near the Altitude of the same Horizontal Line, from whence it fell, decreasing after the same velocity, in which it before accelerated, viz. moving like spaces in proportionable times, accounted from the perpendicular each way, where the two circles unite. Some of our Society judge it to be true, and are laboring to find the demonstration for it," Oldenburg 1965–1986, 4: 571. Wren, Croone, and Hooke were those asked to study the subject; see Birch 1756–1757, 2: 116–117. However, until the meetings of October–November 1668 the problem is presented as one among many others, and one does not get the impression that it is of a particular importance or urgency within the Royal Society. The new phase, beginning in October 1668, is characterized by strong support of some members who were not happy with the way in which such experiments were done in the past.

31. According to Birch, at the meeting of the Royal Society of October 22, 1668, Hooke was asked by the President to tell the Royal Society about his attempt to prove the motion of the earth from observations. Hooke was unwilling to do it and diverted attention from the question concerning the motion of the earth to the question concerning the laws of motion. The proposal was "that the experiments of motion might be prosecuted, thereby to state at last the nature and laws of motion," Birch 1756–1757, 2: 315. The first reaction of Brouncker and the others seemed to have been that such a thing has already been done by Huygens and Wren, so the decision was postponed. In the end, Oldenburg was asked to write to Huygens and Wren for "their speculations and trials of motion." See also A. R. Hall 1966; M. B. Hall 1991. From this account it is clear already that there are camps within the Royal Society: some are unsatisfied with what has been done so far and who do not consider the problem solved, like Hooke, and, others suggest new hypotheses on motion.

32. It is interesting to note that this is exactly the way in which Sorbière describes the activity of the Royal Society: a group of people who are searching in all the books ever written for experiments and hypotheses to be tested in the meetings.

33. Wallis's theory was read before the Society on November 26, 1668, Wren's theory on December 17, and Huygens's account on January 7. See Wallis 1668 and Wren 1668. Huygens's theory was not published in the same issue, however, leading to an exchange of letters between Huygens and Oldenburg on this issue. Huygens's paper was eventually published in the following year. All three papers are reproduced in Chapter 8 of this volume.

34. And this is how Oldenburg presents the whole enterprise to a foreign correspondent: "The Royal Society is now hot upon examining the laws of motion. They are collecting and considering the reflections of some very

acute philosophers upon the topic. Wallis, Wren and Huygens (who are all three Fellows of this Society) have already presented their theories to the Society, and we eagerly await what opinion of them our experiments will teach us," Oldenburg to Sluse, January 26, 1668/9, Oldenburg 1965–1986, 5: 359. In the same letter, however, Oldenburg extends the discussion outside the circle of the Royal Society, informing Sluse that he had asked Vogel to send him Joachim Jungius's manuscript on the theory of motion (*Phoronomica*) for a comparison and maybe for publication.

35. See Oldenburg to Huygens, February 4, 1668/9 (Oldenburg 1965–1986, 5: 371), in which Oldenburg informs Huygens that the Royal Society is still working on the problem of motion, but many are convinced of the truth of his theorems, which are in accord with those of Wren. The accord between Huygens's and Wren's laws of motion is also the declared reason why Oldenburg does not publish Huygens's paper together with the other two, a fact that led to a new controversy between Huygens and Oldenburg.

36. "The quiescent matter hath no resistance to motion (save what it may have from circumstantial incumbrances, or, if there be any innate propensity to the contrary motion, as in gravity is supposed,) I take for granted amongst most of the moderns; & I see nothing to the contrary why I should not be of that opinion," Wallis to Oldenburg, December 3, 1668, Oldenburg 1965–1986, 5: 218.

37. This is, of course, the main problem of Descartes's physics: in what way can motion (a mode of substance) pass from one body to another? On this question, Wallis's answer is again rather ambiguous: "Whether motion passe out of one subject into another, must be first explained; for in a sense it doth, in a sense it doth not. You will see by my hypothesis what I think of it. In summe, in all percussion the body striking looseth of its swiftness, & the other gains if before at rest. If both before were in opposite motion, both loose of their motion & both going from the other in such proportion as is there expressed," Oldenburg 1965–1986, 5: 218.

38. The "global" law of conservation of the total quantity of motion in the world is, of course, one of the main principles of Descartes's physics: a very debated principle in physics and metaphysics alike, since it raises questions regarding God's action into the created world.

39. Wallis, in his letter to Oldenburg of December 1, 1668, tries to show that his theory, although mathematical, is concerned with physics in Descartes's fashion: "This is the clear account of my thoughts as to those Queries ... consonant to that short synopsis of my Doctrine of Motion which I lately sent you. Of which I have this to add in reference to one of your letters in pursuance of it; where you tell mee that *the Society in their present disquisitions have rather an Eye to the Physical Causes of Motion, & the Principles thereof, than the Mathematical Rules of it*. It is this. That the Hypothesis I sent, is indeed of the *Physical* Laws of Motion, but *Mathematically* demonstrated. For I do not take the Physical & Mathematical Hypothesis to contradict one another at all. But what is Physically performed, is Mathematically measured. And there is no other way to determine the Physical Laws of Motion exaclty, but by applying the Mathematical measures & proportions to them," Oldenburg 1965–1986, 5: 221. In a letter from December 9 and 10, Wallis elaborates further the difference between mathematics and physics. We can use geometry for answering questions concerning the phenomena of collisions; mathematics is enough to understand how the collision takes place and what happens with the bodies after the collisions. For the general question concerning the cause of motion and the conservation of the total quantity of motion we need more than mathematics, we need to *postulate*

the existence of matter, motion and a *vis motrix*, Oldenburg 1965–1986, 5: 230–231.

40. One such (unpursued) alternative was the plan to find, read, and maybe publish Jungius's *Phoronomica*, still in manuscript at that date.
41. See, for example, Hooke's *Potentia Restitutiva, or Spring*, published in Hooke 1679. In it Hooke investigates the cause of elasticity, starting from a wide range of experiments, then postulating a number of hypotheses about the nature of bodies and motion. For example, Hooke's definition of "physical bodies" reads: "By Body I mean somewhat receptive and communicative of motion or progression. Nor can I have any other Idea thereof, for neither Extension nor Quantity, hardness nor softness, fluidity nor fixedness, Rarefaction nor Densation are the properties of Body, but of Motion or somewhat moved," Hooke 1679: 7. In other words, Hooke gives a definition of "body" in terms of powers, not extension. In this way, he is closer to the Gassendists than to the Cartesians, in asserting an internal power of particles, independent from their extension or motion: "I do therefore define a sensible Body to be a determinate Space or Extension defended from being penetrated by another, by a power from within," Hooke 1679: 8.
42. The solution of Neile was published, with a demonstration by Wallis, in *De Cycloide*, 1659.
43. Ben Jonson, *What is a Poet?* "A poet is that which by the Greeks is called a maker or feigner: his art of imitation or feining, expressing the life of men in fit measure, numbers and harmony, according to Aristotle from the work poien which signifies to make or to feign," Jonson 1756, 7: 145.
44. In the letter that accompanies the long-awaited hypothesis on motion, classified as "Mr Neile's Principles of Philosophy," Royal Society Classified papers III, i, no. 48. See Hunter 1989 for a discussion.
45. "And since this resistance is opposed to all directions to any external impetus whatever, that internal motion of the most minute particles in every body must exist in an almost endless variety, by which they resist any external impulse, which we have often observed happening with such swiftness that the rebounding of bodies from each other is too quick for the eye, and yet the hardness and coherence of the particles in dense bodies is such that one particle cannot be disengaged from another, but they are as it were bound together and entangled by a wonderful variety of respective motions," Oldenburg 1965–1986, 5: 525
46. "Now if we may allow that such very minute particles may travel more often from some one direction than from any other, and that perhaps with a greater velocity and with fewer or shorter pauses than they did before, it follows that the whole body *A* may be transported according to the aforesaid rules after a very brief interval of time, with periods of rest," Oldenburg 1965–1986, 5: 526.
47. "Now if we suppose the body *A* to be a mass of very minute particles, moving with respect to each other in a great variety of motions and with intervening periods of rest, according to the above-stated rules by which the particles are now brought together and now separated again from each other by a motion which first tends to compress them together, then tends to make them fly apart, it follows that each and every particle is in the most minute spaces of time moved this way and that, upwards and downwards, backwards and forwards and, in a word, every which way (if I may so speak), and yet it is not greatly moved with respect to the neighboring particles," Oldenburg 1965–1986, 5: 525–526.
48. "I thinke nothing but a power that could bring chaos to order can putt a totall stoppe to the motions for the least particle of matter no sooner rests but there may be supposed ten thousand thousand other particles (as it were) to make warre upon it and to thrust it out of its place or at least to cutte of[f]

something from it or if it be at once surrounded on every side it may soon be delivered from that constipation either in whole or in part I can easily imagine particles to stoppe and to continue quiescent if their neighbours would not trouble them but where I presuppose nothing but confusion how to make a durable quiet I think is very difficult," Oldenburg 1965–1986, 5: 543.

BIBLIOGRAPHY

Alexandrescu, V. and Jalobeanu, D., eds. (2003) *Esprits Modernes, Études sur les modèles de pensée alternatifs au XVI–XVIIIe siècles*, Bucharest and Arad: University of Bucharest Press and Vasile Goldis University Press.
Anstey, P. R. (2005) "Experimental versus speculative natural philosophy" in eds. P. R Anstey and J. A. Schuster 2005, pp. 215–242.
Bacon, F. (1620) *Instauratio magna*, London.
Bertoloni Meli, D. (2006) *Thinking with Objects: The Transformation of Mechanics in the Seventeenth Century*, Baltimore: John Hopkins University Press.
Birch, T. (1756–1757) *The History of the Royal Society of London*, 4 vols, London.
Borelli, G. A. (1667) *De vi percussionis*, Bologna.
Boyle, R. (1999–2000) *The Works of Robert Boyle*, 14 vols, eds. M. Hunter and E. B. Davis, London: Pickering and Chatto.
Descartes, R. (1991) *Principles of Philosophy*, trans V. R. Miller and R. P. Miller, Dordrecht: Kluwer.
———. (1996) *Œuvres de Descartes*, revised edn, 11 vols, eds. C. Adam and P. Tannery, Paris: Vrin.
Des Chene, D. (2000) "On laws and ends: A reply to Hattab and Menn," *Perspectives on Science*, 8: 144–163.
———. (1996) *Physiologia: Natural Philosophy in Late Aristotelian and Cartesian Thought*, Ithaca and London: Cornell University Press.
Fichant, M. (1978) "Les concepts fondamentaux de la mecanique selon Leibniz in 1676," *Studia Leibniziana: Supplementa*, 17: 219–237.
———. (1993) "Mechanisme et metaphysique: Le retablissement des formes substantielle," *Philosophie*, 10: 27–60.
Freddoso, A. J. (1991) "God's general concurrence with secondary causes: why conservation is not enough," *Philosophical Perspectives*, 5: 553–585.
Gabbey, A. (1980) "Force and inertia in the seventeenth century: Descartes and Newton" in ed. S. W. Gaukroger 1980, pp. 196–229.
Garber, D. (1992) *Descartes' Metaphysical Physics*, Chicago: University of Chicago Press.
Gaukroger, S. W., ed. (1980a) *Descartes: Philosophy, Mathematics, and Physics*, Sussex: Harvester Press.
———. (1980b) "Descartes' project for a mathematical physics" in ed. S. W. Gaukroger 1980a, pp. 97–140.
———. (1995) *Descartes: An Intellectual Biography*, Oxford: Oxford University Press.
Glanvill, J. (1661) *The Vanity of Dogmatizing*, London.
———. (1665) *Scepsis scientifica: or Confest Ignorance on the Way to Science*, London.
Hall, M. B. (1991) *Promoting Experimental Learning: Experiment and the Royal Society 1660–1727*, Cambridge: Cambridge University Press.
Hall, A. R. (1966) "Mechanics and the Royal Society, 1669–1670," *British Journal for the History of Science*, 3: 24–38.
Hattab, H. (2000) "The problem of secondary causation in Descartes: a response to Des Chene," *Perspectives on Science*, 8: 93–118.

Hearne, T., ed. (1764) *Itinerary of John Leland the Antiquary*, 2nd edn, 9 vols, Oxford.
Hooke, R. (1679) *Lectiones Cutlerianae*, London.
Hunter, M. (1980) "Latitudinarianism and the 'ideology' of early Royal Society" in Hunter 1989, pp. 27–45.
———. (1989) *Establishing the New Science: The Experience of the Early Royal Society*, Woodbridge: Boydell Press.
———. (2007) "Robert Boyle and the early Royal Society: a reciprocal exchange in the making of Baconian science," *British Journal for the History of Science*, 40: 1–23.
Hunter, M. and Wood, P. B. (1986) "Towards Solomon's House: rival strategies for reforming the early Royal Society," *History of Science*, 24: 49–108; reprinted in Hunter 1989, pp. 185–244.
Huygens, C. (1669) "A summary account of the laws of motion," *Philosophical Transactions*, 4: 925–928.
———. (1888–1950) *Oeuvres complètes de Christiaan Huygens*, 22 vols, The Hague.
Jalobeanu, D. (2002) "The two cosmologies of René Descartes" in eds. V. Alexandrescu and D. Jalobeanu 2002, pp. 75–95.
———. (2003) "Le modele mathematique de l'individuation chez Descartes," *ARCHES*, 5: 81–110.
———. (2006a) "The politics of science: strategies of obtaining consensus within the Royal Society," *Zeitsprunge, Forschungen zur Fruher Neuzeit*, 10: 386–400.
———. (2006b) "Bodies, laws and the problem of secondary causation in Descartes' natural philosophy," *Studia Universitatis Babeș-Bolyai*, PHILOSOPHIA, 1, http://hiphi.ubbcluj.ro/studia/?page=nr.php&an=2006&nr=1&lang=en
———. (2008) "Bacon's Brotherhood and its classical sources," *Intersections*, 11: 197–231.
Jonson, B. (1756) *The Works of Ben Jonson*, 7 vols, London.
Joy, L. (1993) *Gassendi the Atomist: Advocate of History in the Age of Science*, Cambridge: Cambridge University Press.
Lennon, T. M. (1993) *The Battle of the Gods and the Giants*, Princeton: Princeton University Press.
Lynch, W. T. (2001) *Solomon's Child: Method in the Early Society of London*, Stanford: Stanford University Press.
Merchant, C. (1967) "The Controversy over Living Force: Leibniz to d'Allembert," PhD Thesis, University of Wisconsin.
Menn, S. (2000) "On Dennis Des Chene's *Physiologia*," *Perspectives on Science*, 8: 119–143.
Oldenburg, H. (1965–1986) *The Correspondence of Henry Oldenburg*, 13 vols, eds. A. R. Hall and M. B. Hall, Madison, Milwaukee, and London: University of Wisconsin Press, Mansell and Taylor & Francis.
Pessin, A. (2003) "Descartes's Nomic Concurrentism: finite causation and divine concurrence," *Journal of the History of Philosophy*, 41: 25–49.
Ronan, C. A. and Hartley, H. (1960) "Sir Paul Neile, F.R.S. (1613–1686)," *Notes and Records of the Royal Society of London*, 15: 159–165.
Schmaltz, T. (2003) "Cartesian causation: body–body interaction, motion, and eternal truths," *Studies in History and Philosophy of Science*, 34: 737–762.
Slowick, E. (2002) "Descartes' quantity of motion: new age holism meets the Cartesian Conservation Principle," *Pacific Philosophical Quarterly*, 80: 178–202.
Sprat, T. (1667) *History of the Royal Society*, London.
———. (1708) *Observations upon Mons. De Sorbiere's Voyage into England, written to Dr. Wren*, London.

Sorbière, S. (1709) *A Voyage to England Containing many Things Relating to the State of Learning*, London.
Syfret, R. H. (1950) "Some early critics of the Royal Society: Stubbe, Crosse, Causabon," *Notes and Records of the Royal Society*, 8: 20–64.
Wallis, J. (1659) *Tractatus duo prior, de cycloide et corporibus inde gentis: posterior, epistolaris in qua agitur de cissoide, et corporibus inde gentis, et de curvarum*, Oxford.
———. (1668) "A Summary account of the General laws of motion, by way of a letter written to the Royal Society," *Philosophical Transactions*, 43: 864–866.
Woolhouse, R. S. (1994) "Descartes and the nature of body," *British Journal for the History of Philosophy*, 2: 19–33.
Wren, C. (1668) "Lex naturæ de Collisione Corporum," *Philosophical Transactions*, 43: 867–868.

7 On Composite Systems
Descartes, Newton, and the Law-Constitutive Approach

Katherine Brading

GENERAL INTRODUCTION

The title of this volume is *Vanishing Matter and the Laws of Motion*. The context is the early modern debate over how best to revise or replace the Aristotelian account of individual bodies as the things of which the world is constituted. In the context of Newtonian mechanics, the phrase "vanishing matter" refers to the view that this theory provided a dynamical account of the behavior of large-scale material bodies, while at the same time treating them as mathematical entities and providing no insight into their nature. There is something right about this: Newtonian mechanics enables us to treat the behavior of bodies without first saying anything about their metaphysical nature. This signals an important shift in the relationship between dynamics and matter theory. However, the phrase "vanishing matter" implies the vanishing of matter theory from physical theory, as though Newtonian mechanics is silent about metaphysical questions concerning the nature of material bodies. I think there is a different way to understand the shift that took place. Far from being silent, Newtonian mechanics has significant implications for matter theory. With hindsight, we can see that it is not that the traditional questions of matter theory vanish, but rather that the development of Newtonian mechanics enriches the logical and philosophical space in which matter theory is to be explored, profoundly changing the framework within which these issues are to be addressed. In particular, various metaphysical aspects of matter theory—such as whether bodies have actual parts—become entangled with (rather than being independent of, and prior to) the details of the physics. This marks a deep change in the relationship between physics and metaphysics, and one to which any later attempt to do matter theory must pay due attention: the philosophical space changes with the advent of Newtonian theory, and there is no going back.

That is a big claim, and very general, so now to specifics. I have one very narrow line of argument that I want to push, and it concerns one way to read the implications of the treatment of bodies in Descartes and in Newton. I will argue that the extension of this treatment to composite systems reveals important consequences for matter theory. I will begin with a brief discussion of bodies in Descartes's system, by way of introduction, and this will enable me to set up the issues concerning composite systems that I want to focus on in this chapter.

SPECIFIC INTRODUCTION

In his *Principles of Philosophy* (1644), Descartes offered his three laws of nature, concerning the behavior of "things" and of "bodies." Here are the laws as he stated them in his *Principles* (Part II, arts. 37, 39, and 40):

> The first law of nature: that each thing, as far as is in its power, always remains in the same state; and that consequently, when it is once moved, it always continues to move.
>
> The second law of nature: that all movement is, of itself, along straight lines; and consequently, bodies which are moving in a circle always tend to move away from the center of the circle which they are describing.
>
> The third law: that a body, on coming in contact with a stronger one, loses none of its motion; but that, upon coming in contact with a weaker one, it loses as much as it transfers to that weaker body.
>
> (Descartes 1991: 59, 60, 61)

But what are the "things" and "bodies" to which these laws apply? If Descartes's laws are to say anything, then there must be bodies to which they refer. Call this the "problem of bodies." For Descartes, the answer is "parts of matter." Famously, however, this answer masks a difficulty that Descartes never satisfactorily resolved. In this section I will briefly review what this difficulty is and how it arises, and outline one possible response, which I call the "law-constitutive" approach.[1] With this in place, I will then turn attention to the main purpose of this chapter: the application of the law-constitutive approach to composite systems.

According to Descartes, on the one hand we have a clear and distinct idea of matter as extended, and on the other hand experience teaches us that this extension is divided into parts, having various shapes and motions. If our metaphysics is to be founded on clear and distinct ideas *and* to include parts of matter, then we had better have a clear and distinct idea of those parts. For *this* to be possible, Descartes must provide *within his metaphysical system* the resources for dividing matter into parts *such that* we can clearly and distinctly perceive that it is so divided.[2] The answer that Descartes appears to give is that *motion* is the principle by which matter is divided into parts. In *Principles* II, art. 25 Descartes gives his definition of "What movement properly speaking is," and then offers an account of the division of indefinite extension into parts or bodies *through* motion:

one body, or one part of matter, is everything that is simultaneously transported. However, motion is itself defined by appeal to the parts of matter. The resulting view is that motion is defined in terms of bodies, but the division of indefinite extension into bodies is achieved through their relative motions. This is, at best, a rather tight circle. Whatever you might think about this, Descartes's next move is to present his laws of motion and, as we have seen, these refer to bodies. The difficulty we are faced with is that we have laws that refer to bodies while not yet having in hand a completed account of bodies.

There are two ways to respond to this difficulty.[3] On the one hand, you might attempt to "complete" the metaphysical account of bodies, providing criteria of individuation and identity that enable a solution to the problem of bodies prior to the specification of the laws of nature. On the other hand, you might suggest that the laws themselves contribute to the solution of the problem of bodies, such that bodies *are*, in part, whatever satisfy the laws. We expand the rather tight circle where motion and body are inter-defined, and thereby hope to turn a vicious circle into a virtuous one. This is what I call a "law-constitutive" approach to the problem of bodies.

I have argued in detail for this approach to the problem of bodies elsewhere (Brading forthcoming), where I also show that the law-constitutive approach was explicitly adopted by Newton, for whom a necessary condition for something to *be* a physical body is that it satisfy the laws. This claim runs at least from "De Gravitatione" (where his account of bodies as impressed shapes in space includes the requirement that these shapes move according to the laws), to drafts made in preparation for the third edition of the *Principia*. I will not argue for this here. Rather, my goal is to extend the law-constitutive approach to the explicit consideration of composite systems. Once again, I think it is helpful to start from Descartes, and then move to Newton. I will say something later about the extent to which I am willing to argue for the law-constitutive approach to composite systems as an interpretation of either Descartes's or Newton's own views, but my main purpose is not exegesis. Rather, my interest is in how the philosophical landscape of matter theory is changed by the philosophical moves that Descartes and Newton make, and my point will be to display some of the rich and far-reaching metaphysical implications of the approach.

The topic of composite systems has two aspects: (1) composite systems constructed from bodies, and (2) the question of whether those bodies themselves should be regarded as composite systems. In each case, there are metaphysical and physical questions that one can ask. With respect to (1), we should distinguish between such metaphysical questions as "*In virtue of what* is the result a *composite system* rather than merely a *collection* of bodies?" or "What is the principle of unity here?" and physical questions such as "What is the glue that binds the bodies together into a composite system?" As regards (2), when we ask about the dividing of bodies and of composite systems into parts (and thus about the status of the bodies themselves), we should distinguish between metaphysical divisibility and mere physical divisibility. In both cases, (1) and (2), my concern is with the metaphysical questions, and not the physical.

In what follows, I begin with the construction aspect (1), first in Descartes and then in Newton. I argue that basically the same principle of unity emerges from both Descartes's and Newton's work, addressing the metaphysical question "*In virtue of what* is the result a *composite system* rather than merely a *collection* of bodies?" The remainder of the chapter discusses the division aspect (2), where my focus is on the actual and potential parts debate. I suggest that neither Descartes nor Newton is best understood as ascribing to either doctrine, but rather that their work marks an important shift in the philosophical framework within which the issue of divisibility should be addressed.

FROM BODIES TO COMPOSITE SYSTEMS IN DESCARTES

I will begin with the construction project (1) as it appears in Descartes and in Newton,[4] and I will argue that the same kind of answer to the metaphysical "in virtue of what" question emerges from both Descartes's and Newton's work.

When considering Descartes's approach to this issue, it is worth starting from the laws of nature that he presents in his manuscript *The World* (c. 1633; Descartes 1998). There are important differences between the cosmological projects set out in *The World* and the *Principles*,[5] and between the two versions of the laws, but I think that the "in virtue of what" question receives essentially the same answer. This answer is immediately evident in *The World*, but is somewhat masked by the changes to the laws that Descartes makes in the *Principles*, and for this reason it is helpful to begin with *The World*.

Descartes begins with a conservation law for the behavior of a lone body, free from collisions with other bodies. He writes:

> The first is that each individual part of matter continues always to be in the same state so long as collision with others does not force it to change that state.
>
> (CSM 1 93)

What we need now is an account of what it is to stay in the same state, of what it is to change state, and also of under what conditions change so defined can take place. Descartes continues as follows:

> That is to say, if the part has some size, it will never become smaller unless others divide it; if it is round or square, it will never change that

shape unless others force it to; if it is brought to rest in some place, it will never leave that place unless others drive it out; and if it has once begun to move, it will always continue with an equal force until others stop or retard it.

(ibid.)

Adopting the law-constitutive approach, a necessary condition for the individuation and identity of a part of matter, or a physical body, is that when it is free from collisions, it retains the same shape, size, and quantity of motion. I have argued elsewhere for this approach to individual bodies as a solution to the problem of bodies in Descartes (Brading forthcoming), and will not do so here. Rather, I will move directly on to the consideration of composite systems.

Having stated his first law, the next step in Descartes's project is to move from the consideration of an isolated individual body to an analysis of what would happen if a second body was added to the conceptual structure. The second law of *The World* reads:

I suppose as a second rule that when one body pushes another it cannot give the other any motion unless it loses as much of its own motion at the same time; nor can it take away any of the other's motion unless its own is increased by as much.

(CSM 1 94)

This is a law of conservation of the total quantity of motion of a composite system: it extends the first law from the single body case to the case of a pair of bodies, and provides us with a law for a composite system of colliding bodies considered as isolated from the rest of matter.[6]

At least, that is how I think we should read it. Interpreting the second law as a conservation law follows one of the two main lines of interpretation in contemporary literature.[7] The other standard interpretation treats the second law as a law of impact, judging it by its success at determining the outcome of collisions.[8] But this law is not sufficient to determine the outcome of a collision because it does not determine how the total quantity of motion will be distributed among the component bodies after the collision and it says nothing about the subsequent directions of these component bodies. Viewed in this way, the law is a failure.[9] However, viewed as a conservation law it achieves a very important goal, by generalizing the first law for single bodies to the case of a pair of interacting bodies: a composite system (not subject to collisions from outside) satisfies conservation of quantity of motion just as a lone body (not subject to collisions from outside) does. This is a global claim about the composite system as a whole, and not a claim about its parts.[10]

Our task now is to adopt the law-constitutive approach with respect to Descartes's laws, as presented in *The World*, and to examine the implications (if any) for the metaphysics of composite systems. It seems to me that we can draw three important conclusions:

- First, just as the first law gives a necessary condition for the individuation and identity of bodies, the second law generalizes this condition to composite systems.
- Second, satisfaction of the second law is (partially) constitutive of what it is to be a composite system: when it is free from outside collisions, is conserves its total quantity of motion. It is a composite because there were two bodies initially, and it is a whole because the composite satisfies the conservation law. Thus, the second law provides a *principle of unity* in virtue of which the composite is a genuine whole.
- Third (and this is a negative conclusion), the laws of nature offered by Descartes in *The World* cannot be used to individuate the component bodies of a composite system. The second law is silent on the behavior of the components considered individually, and while the third law ascribes a tendency to the components, this is not sufficient to determine the behavior of the components within the composite system. The upshot is that, on this approach, there *are* no determinate components.

The positive proposal here is that the laws offer a principle of unity in virtue of which a composite forms a genuine whole as opposed to a mere collection. Specifically, the composite conserves its total quantity of motion, and the claim is that this is a necessary and sufficient ground for a genuine unity.[11] However, on the negative side, this genuine unity lacks determinate components.

The failure to determine the redistribution of the quantity of motion among the component bodies following collision is something that Descartes seeks to address in his revision of the law in the *Principles* (where it appears as the third law, see above) and the accompanying rules of collisions. This law, unlike that appearing in *The World*, is directed at the behavior of the parts of a composite system. Together with the rules of collision (*Principles* II, arts. 46–52), it seeks to determine how the motion of the component bodies of is affected by a collision. Note that this determination remains subject to the *global* constraint on the composite system as a whole that the total quantity of motion of the whole remain unchanged.[12] The first law of the *Principles* remains essentially the same as that of *The World*, for our purposes, and the old third law of *The World* now becomes the second law of the *Principles*.

Viewed from the law-constitutive perspective, we can say that the *Principles* attempts a significant step forward: in addition to providing a principle of unity for composite systems free from outside collisions, the third law and the rules of collision can also be used to try to determine the behavior of the component bodies of a composite system, and therefore to provide a necessary condition for the individuation and identity of the components: a necessary condition for something to *be* a component body of a composite system is that it move according to the third law and the rules of collision.

By now we have moved far from Descartes exegesis: throughout his statement of the laws and the rules of collision, Descartes writes as if the bodies that are their subject matter are already given. I have said that Descartes has not, in fact, succeeded in providing the bodies that are the subject matter of his laws, prior to his statement of the laws, and I will come back to this point later on (see 'From bodies to their parts', below). What I am doing here is adopting one possible solution to this problem and re-interpreting Descartes's laws in this light with a view to displaying the philosophical consequences for matter theory. Viewed through the lens of this law-constitutive approach, the following points emerge:

- First, the laws of Descartes's *Principles* provide (or attempt to provide) necessary conditions for the individuation and identity of not just isolated bodies, but also isolated composite systems, and the component parts of those systems.
- Second, the third law provides a principle of unity for composite systems: a composite system is a unified whole *in virtue of* conserving its total quantity of motion.
- Third, insofar as the third law and the rules of collision fail to solve the problem of collisions, they are also insufficient to determine the component bodies of a composite system, and we are still left with a composite system consisting of indeterminate parts.
- Fourth, as regards the conditions placed on them by the laws, the *ontological status* of isolated bodies, isolated composite systems, and component parts of isolated composite systems are much on a par with one another. This is a point we shall return to later when we consider the status of the parts of bodies. Descartes has available a criterion for answering the metaphysical question "In virtue of what is a given entity a genuine unity?" for isolated bodies and composite systems, including the universe as a whole, and that answer is "in virtue of possessing a constant total quantity of motion." An advantage of this approach is that we get a unified approach to individuation through conservation of total quantity of motion, all the way up to the cosmos as a whole.

With this application of the law-constitutive approach to Descartes's system in mind, I want to turn our attention now to Newton, and to his construction of composite systems, during which I will draw some conclusions for matter theory.

FROM BODIES TO COMPOSITE SYSTEMS IN NEWTON

I am going to start from the assumption that Newton explicitly proposed a version of the law-constitutive view according to which a necessary condition for an entity to be a physical body is that it satisfy the laws of motion. As noted above, I have argued for this elsewhere and my goal here is to extend this approach to composite systems.

Just as for Descartes, Newton's first law concerns the behavior of a single isolated body. The question then arises: how does Newton progress from the motion of a single isolated body to the behavior of interacting bodies?

Newton's general strategy in the *Principia* is exactly that found in Descartes: we proceed by construction from the behavior of isolated individuals to the behavior of composite systems via conservation laws. But in Newton the strategy is implemented with clear success when it comes to the component parts of composite systems. From the beginning of his consideration of individual bodies, Newton is interested in saying precisely how the state of a body changes as a result of a collision. Newton's second law tells us in what way a body's state will be changed by the action of an external force, and, crucially, this change is quantifiable. It is the third law, however, that allows Newton to extend his analysis to the behavior of bodies interacting with one another. By means of his third law, Newton achieves an answer to the distribution question and an extension of the conservation of the linearity of motion from single bodies to composite systems;[13] his solution provides a rule that determines uniquely and quantifiably the outcome of two body collisions and interactions.[14]

From the law-constitutive perspective, this is important not just because it solves a problem in mechanics, the problem of collisions, but, more fundamentally, because it extends the law-constitutive approach to the *component bodies* of a composite system. Putting the point more dramatically: it gives necessary conditions for something to *be* a part of a composite system, *and sufficient conditions for those parts to be determinate*. Thus, this solves a problem in physics, but also—when viewed from the law-constitutive perspective—a problem in metaphysics.

If we adopt the law-constitutive approach, we can draw the following conclusions:

- First, Newton's laws provide necessary conditions for the individuation and identity of bodies, composite systems, and the component parts of those systems.
- Second, the laws provide a principle of unity for composite systems. The role of the third law is to determine the behavior of component bodies of a system, behavior that must be consistent with the first law continuing to hold for the composite interacting system as a whole. In other words, an analogous principle of unity for composite systems that we drew from Descartes's system is also available in Newton's system: conservation of quantity and direction of motion of the whole, when free from external interactions.
- Third, the laws are sufficient for the parts of a composite system to be determinate.
- Fourth, we note—as we did when considering Descartes's system—that the ontological status of isolated bodies, isolated composite systems, and component parts of composite systems is equal. This approach does not deliver any account of ontological priority of bodies over systems, or vice versa. Rather, they are all on a par.

As I stressed at the outset, my goal is not Newton exegesis but rather the question of how best to think about bodies and composite systems in the light of the legacy left to us by Descartes and Newton. But let me be clear about how far I am willing to support what I have said as exegetical. First, I *do* think that the law-constitutive approach to bodies is explicit in Newton, as I argue in Brading forthcoming. Second, I *do* think the constructional strategy for how to build composite systems out of bodies is explicit in Newton. The argument for this is set out in the Appendix, both with respect to how Newton presents his theoretical system and also with respect to how he applies it. Finally, while I do *not* think that the law-constitutive approach to composite systems is explicit in Newton, I *do* think it follows very naturally from the conjunction of the law-constitutive approach to bodies plus the constructional strategy, both of which I maintain are explicit in Newton.

THE TRANSFORMATION OF MATTER THEORY

My claim is that the law-constitutive approach to the construction of composite systems from bodies leads to important metaphysical results traditionally associated with matter theory. First and foremost, it provides a principle of unity in virtue of which a composite system constitutes a genuine whole rather than a mere collection. This principle of unity is not about merely *physical* unity. It is not, for example, about the glue that binds a composite system together (for this, on the Newtonian picture, we need specific force laws). Moreover, the unity of the bodies from which the composite is made is itself grounded in the very same principle. The conservation of quantity of motion by a body, or by a composite system, should be read as a *metaphysical* principle, the necessary and sufficient ground of the unity of the body or system. This proposal for a principle of unity can be challenged, of course, but it should be challenged as a metaphysical claim about matter theory, and thus duly recognized as such.

In the light of this, what should we say about the apparent absence of matter theory in Newtonian mechanics? I think we have an alternative account of why Newtonian mechanics appears to be silent about matter theory. It is not simply that bodies are being treated mathematically, and that this can be done without *first* providing a theory of matter. This is true, but it is not the whole story. It is *not* that we do not *have* to provide a matter theory first, it is that we *cannot*: on the law-constitutive approach, the matter theory comes along with the laws. The laws give necessary conditions on *what it is* to be a body, and on *what it is* to be a composite system of bodies. Furthermore, as I shall now argue, the laws give necessary conditions on *what it is* to be a *part* of a body. Traditionally, these questions belong to matter theory and to metaphysics, but with the development of Newtonian mechanics I think that the two become entangled. It is not that matter theory vanishes, but that it is no longer prior to mechanics. In the next section, below, we will see how this plays out when it comes to the debate over the status of the parts of bodies.

Clearly intertwined with the story I have told is the search for the laws of collision. In her chapter in this volume, Jalobeanu returns to the developments that took place historically between the proposals of Descartes and Newton, focusing on the largely forgotten contributions of William Neile. At the time, the challenge posed by the problem of collisions was seen as twofold: (1) to search for the properties by which to characterize bodies so that the problem of collisions can be solved, and (2) to "account for" those properties in terms of something else. This "accounting for," on the Cartesian model, was the reduction of the dynamical properties (such as hardness and elasticity, for example) to the geometrical properties of size, shape, and motion (this being what Jalobeanu calls the "strong program" of Cartesian geometrical reductionism). Thus, according to Jalobeanu, Neile repeatedly expresses his concern that the problem of collisions is not adequately solved until we have given a definition of the basic concepts used in our laws, including "hardness" and "elasticity," *prior to and independently of* our specification of the laws: matter theory is prior to physics.

Jalobeanu's chapter beautifully illustrates the tension between the inherited view that matter theory is prior to physics (such as is exemplified by the Cartesian "strong program") and the newly emerging law-constitutive approach. It seems to me that we can see the protagonists in the collisions debate wrestling with this very issue. The question is: if we have a solution to (1), what more could we possibly want? More precisely, what is it that we asking for in (2), and what work can this "something more" be made to do?[15] The lesson we should take away is: nothing. A *complete* characterization of the properties of bodies can, in principle, be given by the laws: there are no "residual" questions that a separate matter theory should address; matter theory is absorbed into physics. This is a profound shift in the relationship between physics and metaphysics, the seeds of which were sown by Descartes, and which forms part of our inheritance from Newtonian natural philosophy. As Murray, Harper, and Wilson argue (Ch. 8 of this volume), Newton makes the task of answering (1) an empirical matter; I have argued here that he also renders question (2) a nonquestion.[16]

FROM BODIES TO THEIR PARTS

In this section we consider the status of the parts of bodies. The composite systems we have considered above are constructed from bodies, and as such have actual parts. The question we will consider here is whether the bodies themselves have actual parts, and if so, what account we should give of those parts. For this purpose, I will frame the discussion in terms of the actual/potential parts debate. Here, I am deeply indebted to Thomas Holden's book, *The Architecture of Matter*, which is all about the actual/potential parts debate in the seventeenth and eighteenth centuries. The actual parts doctrine (see Holden 2004: 80) states that the parts into which a material body can be metaphysically divided (i.e., the parts into which God could break it, even if no natural process could) are *actual* parts, where *actual* parts are parts that are independent existents that exist prior to any act of division, and are ontologically prior to the whole. Thus, given the actual parts doctrine, bodies are composite entities whose parts have a more fundamental ontological status than the bodies themselves. The potential parts doctrine (see Holden 2004: 79), by contrast, states that the parts into which a material body can be metaphysically divided are *potential* parts, where *potential* parts are merely possible existents until actualized by an act of division. As Holden is at pains to emphasize, a crucial issue in the debate concerns the apparent conflict between the infinite divisibility of matter and the actual parts doctrine: conjoined, these two theses imply that every body is constituted by an actual infinity of parts, and this was held by most of those involved in the debate at the time to be seriously problematic.

According to Holden, both Descartes and Newton are, in different ways, actual parts theorists. He writes:

> The actual parts doctrine is quite orthodox in this dominant tradition in early modern physics and metaphysics, and is ratified by nearly all the new philosophers of this period. . . . First, the doctrine is endorsed by philosophers representative of the two great systems within the new science: the system of Descartes and the Cartesians on the one hand, and the system of Newton and the Newtonians on the other.
> (Holden 2004: 86)

The difference between them lies in how they respond to the threat of paradox arising from infinite metaphysical divisibility: according to Holden, while Newton denies infinite metaphysical divisibility, Descartes endorses both the actual parts and the infinite metaphysical divisibility theses while admitting that it is difficult to understand how they fit together.

In the following sections of the chapter, my goal is to do two things. First, I will call into question the claim that Descartes and Newton were actual parts theorists, and I will suggest that neither philosopher's position fits neatly into either the actual or the potential parts camps. Second, I will argue that the law-constitutive approach as applied to the parts of bodies yields an interesting alternative account, and I will suggest that it is one that fits each philosopher much better.

DESCARTES AND THE DOCTRINES OF ACTUAL AND POTENTIAL PARTS

Let us begin by considering the evidence that Descartes subscribes to the actual parts doctrine. First, note that Descartes is clearly committed to the infinite metaphysical divisibility of matter, which he argues for in his rejection of atomism as follows (*Principles* II, art. 20):

> We can also easily understand that it is not possible for any atoms, or parts of matter which are by their own nature indivisible, to exist. The reason is that if there were any such things, they would necessarily have to be extended, no matter how tiny they are imagined to be. We can, therefore, still conceive of them being divided into two or more smaller ones, and thus we know that they are divisible.
>
> (Descartes 1991: 48–49)

Now the question is whether this infinite (or indefinite) divisibility is associated with actual parts, or merely with potential parts. The evidence for the actual parts interpretation offered by Holden relies entirely on Descartes's application of the real distinction to parts of matter. As Holden rightly asserts, Descartes is insistent that the parts of matter are really distinct from one another. For example, in the *Principles* (I, art. 60) Descartes writes:

> For example, from the sole fact that we now have the idea of an extended or corporeal substance (although we do not yet know with certainty that any such substance truly exists), we are however certain that it can exist; and that if it exists, each part of it delimited by our mind is really distinct from the other parts of the same substance.
>
> (Descartes 1991: 27)

However, as Holden remarks, the actual parts interpretation seems to be in conflict with the account of parts of matter that Descartes gives a little later (*Principles* II, art. 25) where he states:

> By *one body*, or *one part of matter*, I here understand everything which is simultaneously transported; even though this may be composed of many parts which have other movements among themselves.
>
> (Descartes 1991: 51)

Holden resolves this apparent conflict by distinguishing between physical bodies (*merely* physically unified beings) and metaphysical "really distinct" individuals. He writes:

> It is true that, for the purposes of his dynamics, Descartes holds that the parts of matter are individuated by their relative motion, such that the rupture and separation of a previously undifferentiated portion of matter creates two distinct bodies from one. . . . And this may seem less like an actual parts account and more like a potential parts analysis where division creates rather than unveils parts. But this account applies merely to dynamics. At the metaphysical level—the level of individuation into "really distinct" substances or independent beings, rather than the merely physically unified beings that concern dynamics—Descartes consistently maintains an actual parts account.
>
> (Holden 2004: 86)

This is Holden's case for Descartes as an actual parts theorist. However, continuing a debate that goes back to Descartes's earliest commentators, the contemporary literature remains divided, both on the issue of whether Descartes endorsed an actual parts metaphysics, and on the relationship between the bodies that are the subject of Descartes's laws and the parts of matter that are the subject of his metaphysics. For example, Normore (2008) endorses the actual parts interpretation, as does Rozemond (2008: 169), who also notes that this interpretation is not uncontroversial, while Lennon (2007) argues that, according to Descartes, the division of extended substance into bodies is mind-dependent. That this is such a thorny area of interpretation suggests that something rather different may be going on, which reconceives the issues not in terms of the traditional actual/potential parts dichotomy.

The lack of consensus in the current literature derives in part from the paucity of quotations in Descartes's corpus directly endorsing either the actual or the potential parts position. Advocates of one or other position attempt to construct an argument that derives their preferred interpretation from premises that Descartes explicitly endorses. It seems to me that the lack of direct evidence is revealing, indicating that we should not try to push Descartes's position into the framework of the actual/potential parts dichotomy. Descartes does not fit neatly into either camp and has, I think, at least the beginnings of a much richer and more original position. As a way to illustrate this suggestion, consider the following passage from the *Principles* that might be taken to support the actual parts interpretation.

Principles II, art. 34 includes in its title that "matter is divisible into an indefinite number of parts."[17] Descartes has been discussing the division of matter into parts by motion, and in this paragraph he argues for "a division of certain parts of matter to infinity"—that is, an *actual* division. The argument considers an ever-restricting neck through which the parts of matter must pass, in making their circular motion (the accompanying diagram is of *non*concentric circles, with the parts of matter setting out from G and heading toward the "neck" at E), and runs as follows:

> For it is not possible for the matter which now fills the space G to fill successively all the spaces of very gradually decreasing size which are between G and E, unless some of those parts adapt their shape and divide as necessary to fit exactly into the innumerable dimensions of those spaces. In order for this to occur, all the particles into which one can imagine such a unit of matter to be divisible, which are truly innumerable, must move slightly with respect to one another; and however slight this movement, it is nevertheless a true division.
>
> (Descartes 1991: 57)

As Roux (2000, esp. pp. 223–230) rightly insists, Descartes is arguing for an actual division of parts to infinity. This is consistent with what

Holden says about Descartes recognizing that his own system must face the challenge posed by actual indefinitely divided matter. However, the above argument does not require that matter *per se* is indefinitely divided. Rather, it requires that there are *some* parts of matter that are actually indefinitely divided, for some periods of time. The paragraph that follows (*Principles* II, art. 35) makes clear that this is Descartes's intention. He writes: "It must be observed that I am not talking here about all matter, but only about some part of it," and goes on to describe larger parts of matter "mingled with" those that are indefinitely divided. Indeed, this coheres well with the various passages where Descartes appeals to the different-sized parts of matter. It seems to me that this passage is one place where Descartes's position can be understood as resisting the potential/actual parts dichotomy.

It is beyond the scope and aims of this chapter to develop a detailed interpretation of Descartes's position along these lines; however, in the final section of this chapter I will argue that the law-constitutive approach gives us a principled way to understand the new approach to the parts of bodies that is, I suggest, embryonic in Descartes.

NEWTON AND THE DOCTRINES OF ACTUAL AND POTENTIAL PARTS

Holden makes frequent claims about Newton being an actual parts theorist, but when it comes to quotations, the evidence is scant, and almost entirely from very early writings. Of the four arguments for the actual parts doctrine that Holden identifies, he finds evidence of only two of them in Newton, both appearing in the so-called *Trinity Notebook*. Indeed, in a footnote Holden himself remarks: "I cannot find an explicit statement of the doctrine in the mature Newton's published writings, though it is strongly suggested in a draft written around the period of the *Principia Mathematica* 2nd edn." His reference is to McGuire 1978: 117, where Newton is discussing the nature of space, and that it has no parts. During this, Newton uses the phrase "nor are there more parts in the totality of space than there are in any place which the very least body of all occupies," and it must be this to which Holden is referring. However, Newton is using this phrase, "least body of all," to illustrate the nature of space; if this is the best evidence that the mature Newton was an actual parts theorist then it is weak evidence indeed.

I suggest that perhaps this is because the dispute becomes a nondispute for Newton, and does so in part because of his law-constitutive approach to bodies (which emerged later than the *Trinity Notebook*; there is no hint of it there). Following the *Principia*, we are hard pressed to find among Newton's metaphysical commitments any that are unrevisable in the light of subsequent developments (arrived at via his maturing methodology for natural philosophy).[18] This is surely true of his atomic hypotheses. Once Newton has developed the law-constitutive approach, the solution to whether there are ultimate bodies becomes dependent on the laws. This point should be clearer after we have considered explicitly the law-constitutive approach to the parts of bodies, to which we now turn.

THE LAW-CONSTITUTIVE APPROACH TO THE PARTS OF BODIES

When discussing the construction of composite systems from bodies above, we saw that, according to the law-constitutive approach, the components of a composite system are those parts of the system that obey the laws. If we now ask of a given body, "what are its component parts?," then the very same law-constitutive analysis of parts can be applied.

I want to stress that this is a metaphysical thesis about the status of parts. Common to the actual and potential parts doctrine is the claim that "a body *is* an aggregate of parts" (be they actual or potential). In the law-constitutive view we say, "a body *is* whatever satisfies the laws," and this is independent of what we may say about actual/potential parts, so we free ourselves from having to make prior metaphysical commitments concerning parts in order to say what a body is. However, the law-constitutive approach *does* make commitments concerning the parts of bodies: to be an actual part of a body is to interact in accordance with the laws. The law-constitutive view denies that *any* old part of a body is also a body, and gives us a rule for telling *which* parts of a body are in fact also bodies. And these bodies are its actual parts.

Let me emphasize that this is not primarily about physical divisibility or about material structure: it is about metaphysical divisibility. Whether a body has actual parts depends on whether it is a composite of parts that themselves satisfy the laws, but such parts need not be *physically* divisible from one another. For example, if the strength of the force dominating the interactions between the parts is great enough, and goes up exponentially with distance, then arguably the parts are not physically divisible from one another, even though they remain metaphysically divisible. As regards material structure, a body might be entirely homogeneous, and yet through how it changes shape over time, or how it moves, might reveal that it has component parts.

My claim is that what is being offered here is an alternative to the dichotomy of actual versus potential parts, and is distinct from these two positions in the following ways. First, the law-constitutive approach to the parts of bodies rejects:

- the actual parts view that *any* part of a body is an actual part;
- the potential parts view that there are *no* actual parts until the body is physically divided into those parts.

Second, the law-constitutive view also rejects:

- the actual parts view that the parts are ontologically prior to the whole;
- the potential parts view that the whole is ontologically prior to the parts.

Elaborating on this second point, the actual parts, as bodies, will have the *same ontological status* as the compound body. A body *may* have actual parts in the law-constitutive view, but those parts (as bodies) will have the *same status* as the compound body has—no body is ontologically parasitical on its parts, and no part is ontologically derivative on the whole. Insofar as the laws contribute to saying what a body *is*, and what a composite system *is*, and what a part of a system *is*, composite systems are just as fundamental as non-composite systems. As a metaphysical thesis, the law-constitutive approach to bodies, component systems, and parts, flattens out the ontological hierarchy.

I offer the law-constitutive approach as a way of thinking both about Descartes's and Newton's own positions on the status of bodies and their parts, and also about how—with hindsight—the actual/potential parts debate came to be dissolved rather than resolved, by the advent of a new position. We are the inheritors of this new position, and it is one to which contemporary metaphysics should pay heed.

In thinking about Descartes's position on the status of bodies and their parts (discussed above), I claimed that for Descartes, all matter is *potentially* divisible *ad infinitum*, that the principle for the *actual* division of matter into parts is motion, and that this actual division is finite in some regions, whereas in others the indefinite divisibility is actualized. This position does not correspond well with either the potential or the actual parts doctrines, but it does fit neatly with the analysis of actual and potential parts that follows from the law-constitutive approach. When it comes to Newton's position, his silence in his mature work on the actual/potential parts issue shouts loudly, to my ears at least, but leaves us with no explicit evidence. Clearly, the issue becomes a nonissue, and my speculation is that this is, at least in part, because of his use of *laws* to guide his search for answers concerning ontology. In particular, Newton is explicit in his law-constitutive approach to bodies, and if he conceives of the parts of bodies as themselves bodies (as perhaps his rules of reasoning might encourage us to believe), then the law-constitutive approach extends to cover the question of divisibility. This is not to say that either Descartes or Newton explicitly advocated such an approach, of course, but it is to claim that—with the benefit of hindsight—we should see in each of Descartes's and Newton's work a crucial chapter in the story of a profound philosophical transformation in the basic framework of matter theory.

CONCLUSIONS

I have sought to challenge the view that Newtonian mechanics deals with "vanishing bodies," treating them as mathematical entities and remaining silent about metaphysical questions concerning their nature. I have argued that if we adopt the law-constitutive approach a different picture emerges, one in which the apparent silence is because matter theory is no longer prior to mechanics, and must be developed in partnership with mechanics. The laws give necessary conditions on *what it is* to be a body, on *what it is* to be a composite system of bodies, and on *what it is* to be a *part* of a body. Traditionally, these questions belong to matter theory, and to metaphysics, but with the development of Newtonian mechanics the two become entangled.

I have offered two examples of the deep matter-theoretic significance of the law-constitutive approach with respect to composite systems. I have argued that if we adopt the law-constitutive approach, Newtonian mechanics provides a principle of unity for composite systems. This principle of unity should be interpreted as a metaphysical principle, providing the necessary and sufficient ground for the composite system to constitute a genuine whole. I have also argued that if we adopt the law-constitutive approach, Newtonian mechanics provides a new position in the actual/potential parts debate, cutting across the traditional dichotomy and offering a new way to approach the question of metaphysical divisibility. Both these aspects of Newtonian mechanics deserve to be treated as philosophically serious contributions to metaphysics, and when this is done they profoundly alter the framework within which discussions of matter theory should take place. I think this shows that Newtonian mechanics is not at all silent when it comes to matter theory.

ACKNOWLEDGMENTS

Material from this chapter was presented at the Bucharest colloquium on "Vanishing bodies and the birth of modern physics" in June–July 2008, and I am grateful to those present for their suggestions. Particular thanks to the editors of this volume for their comments.

APPENDIX: THE CONSTRUCTIONAL STRATEGY IN DETAIL[19]

The Theoretical Solution

It is worth spending some time looking at Newton's constructional strategy in more detail, to see the way in which the strategy allows one to treat any composite system as a body, and also to provide the parts of the composite systems and treat them also as bodies.

The first use of Newton's third law in the *Principia* is found in Corollary III to the laws of motion, where Newton uses his third law to demonstrate that the total quantity of motion before and after a collision between two bodies is conserved. He is demonstrating conservation of motion for two colliding bodies, but he is also doing more than this. For Newton, quantity of motion is not Descartes's scalar notion but rather the vectorial concept, momentum. Unlike Descartes's concept, this concept in conjunction with third law allows us to go beyond the claim that the total quantity of motion is conserved to the redistribution of the total quantity of motion, both in terms of the magnitude of the momentum and in terms of the direction of the motion. We have a quantified solution to the distribution problem, and an extension of the conservation of linearity of motion from single bodies to pairs of colliding bodies. Given two bodies individuated via the numbers attaching to certain quantities (mass and velocity) that then collide, we can now re-identify each of them after the collision because we have a rule for how those numbers change for each individual as a result of the collision.

The next challenge is to generalize this to many-bodied systems. In Corollary IV Newton shows that redistribution of motion in interactions by means of his third law is consistent with first law holding for a composite system treated as a single body via the center-of-mass of the system.[20] The structure of Newton's argument is to build up from the behavior of a set of mutually isolated bodies, via a pair of interacting bodies, to a many-bodies system of interacting bodies. In detail, Newton begins with a set of bodies each of which is freely moving and straightforwardly argues that "the common center of gravity of any two either is at rest or moves forward uniformly in a straight line" (Newton 1999: 422). Then, he considers an isolated system of two interacting bodies. Given the second and third laws, any change in the momentum of one body will be accompanied by an equal and opposite change in the momentum of the other, and hence the center-of-mass of the two-body system remains at rest or in uniform motion.[21]

Next, he adds to this pair of interacting bodies the remainder of the set of mutually isolated bodies with which he began. Combining the above results for the set of non-interacting bodies and the pair of interacting bodies, he concludes that the motion of the center-of-mass of the combination will be unaffected by the interaction of the pair.

Finally, we need to extend this to composite systems in which three or more bodies are interacting. Newton says: "Moreover, in such a system all the actions of bodies upon one another either occur between two bodies or are compounded of such actions between two bodies" (Newton 1999: 20). This is the point that is crucial for the problem of individuation of the component bodies of a many-bodies system. It means that the solution given above for a two-body system holds even when we add more bodies to our system; we are still able to use the rules Newton has given us to calculate the numerical change in velocity that an individual body will undergo as a result of a collision with another body.

On the question of the generalized conservation law, from here Newton concludes that: "Therefore, the law is the same for a system of several bodies as for a single body with respect to perseverance in a state of motion or of rest" (Newton 1999: 423). Conservation of linear momentum is shown to hold for a composite isolated system of interacting bodies via redistribution of motion according to his third law, and the method is to generalize by construction from a single isolated body to a composite isolated system. In this way, we see that the new cosmology is built from isolated subsystems that preserve their state unless acted upon by a force, and that preserve their identity when interacting with other systems, by means of conservation principles.

Newton's Constructional Strategy in Practice

This method of building the cosmos is put into practice in Newton's discussion of planetary motion.[22] For example, in discussing the motion of the satellites of planets Newton (*Principia*, Book 3, Proposition XXII, Theorem XVIII) writes that they will move around their planet but that this motion will be disturbed from a perfect ellipse by the influence of the sun. We can construct the actual motion of a planetary satellite by beginning from a consideration of the satellite plus its planet as a two-body composite system isolated from all other influences.

Newton then goes on to describe the way in which the moon deviates from an elliptical orbit of the Earth, and in Proposition XXV of Book 3 (Newton 1999: 839) he shows how to "find the forces of the sun that perturb the motions of the moon" by considering a system consisting of the moon and Earth only, and then analyzing the actual motion of the moon as a deviation from this idealization.

We end by noting one final feature of this constructional strategy. We have seen that according to Newton the behavior of the three-body system can be analyzed in terms of how the two-body system would have behaved plus a disturbing factor. In other words, the interaction between the sun and the Earth is completely blind to whether or not the moon is present. The overall behavior of the Earth results from its own behavior as an isolated system, plus the contribution arising from its interaction with the sun, plus the contribution from its interaction with the moon, and so forth, and each of these contributions is completely unaffected by whether or not the other contributions are present. In this way, we can proceed to reconstruct the entire universe, adding one body at a time, and nothing that we add will ever require us to go back and recalculate how the sun and the Earth interact.

In conclusion, then, at the heart of the Newtonian cosmos of the *Principia* lies Newton's solution to Descartes's problem of the individuation of material bodies, many crucial aspects of which (taking isolated individual bodies as the starting point, the concept of the state of the body specified numerically and without appeal to the "underlying nature of matter," conservation laws, and the constructional strategy) are found also in Descartes's own solution. Newton certainly made important changes in the process of arriving at his solution, but the basic strategy remains the same.

NOTES

1. The material in this section summarizes claims made in Brading forthcoming. Discussion of the "problem of bodies" can be found within the broader context of the debate over the status of the parts of matter in Descartes's metaphysics, to which I will return below (and see below for references to this literature).
2. Descartes's God is so powerful that he could divide matter into parts in ways incomprehensible to us, presumably, but that will not do here because Descartes requires that we clearly and distinctly perceive that matter is so divided. Therefore, on Descartes's own terms, God must be dividing matter into parts in a way that is intelligible to us and can be accounted for within Descartes's metaphysical system. I think that the issue of our clear and distinct perception that matter is so divided poses a *prima facie* challenge to Normore's recent (and intriguing) suggestion that Descartes takes the individuation of the parts of matter as basic (see Normore 2008), and similarly that those who distinguish Descartes's "parts of matter" from the bodies that are the subject of his physics (such as Holden 2004) owe us an account of this division into parts that satisfies the clarity and distinctness requirement.
3. Normore (2008) has recently suggested that there may be a third option: that we take individuation of the parts of matter as basic. He makes a strong case for this suggestion but, as noted above, I would like to know how he responds to the requirement that our perception of the division of extended matter into parts be clear and distinct.
4. The sections addressing (1) draw heavily on joint work with Dana Jalobeanu, friend and long-term collaborator.
5. Jalobeanu (2003) argues for deeper differences between the two projects than has been hitherto acknowledged in the literature.
6. Garber (1992) points out that although in *The World* and the *Principles of Philosophy* the law of conservation of quantity of motion is presented as a special case of the more general principle that a system will conserve its state unless acted upon externally, chronologically Descartes had the special case first and the general case appears for the first time in *The World*.
7. Gaukroger (1995), for example, views both the first and second laws of *The World* as conservation laws.
8. Garber (1992), for example, calls this law, and its development in the *Principles*, the "law of impact."
9. Indeed, having labeled the second law the "law of impact," Garber goes on to criticize the law for failing to solve the problem of collisions, concluding: "Descartes's purported impact law in *The World* is, thus, no impact law at all" (1992: 232). It seems to me that this counts heavily against the "impact" interpretation: there is a natural interpretation that renders the law successful (the conservation law approach) and surely, all things being equal, this is to be preferred over an interpretation whose outcome is that the law is obviously a failure.
10. The eventual target is the indefinitely extended cosmos in which motion is constantly redistributed in accordance with the general principle that the total quantity of motion of the universe as a whole is conserved.
11. Descartes himself never interpreted his law in this way, of course. However, Spinoza's approach suggests that it is not a huge leap to interpreting the conservation law for composite systems as a principle of unity.
12. Gabbey 1980 and Garber 1992 have argued that this version of the law is best viewed *not* as a conservation law, but as a law about collisions based on

the idea that a collision is a contest between the two bodies. Garber writes: "... the impact law, law B of *The World*, appears as law 3 of the *Principles*, considerably changed from its initial statement. The contest view, at best implicit in the earlier discussion, becomes the heart of the law, now clearly distinguished from the conservation principle...." (1992: 234–235). For more on this contest view of forces see Gueroult 1980. A better interpretation, in my opinion, is that the law *remains* a conservation law for the composite system as a whole, but the problem of redistribution is now tackled in terms of a contest. Gaukroger 2000 has offered a distinct and powerful interpretation of the approach Descartes takes to the problem of redistribution, arguing that Descartes is using the model of statics, and in particular a balance, to work out the rules of collision. Thus, a lighter body will never raise a heavier body placed at an equal distance from the pivot point. The "balance" account is made even more convincing by the fact that Wren and Huygens both used balance analogies in their attempts to solve the problem of collisions (in response to the Royal Society challenge); see Radelet 2000.

13. The genesis of this solution can be traced in the way Newton's laws develop through earlier manuscripts to their final incarnation in the *Principia*. For discussion of the development of the laws see especially Westfall 1971: 439ff, and Herivel 1965.

14. A case for this view of the role of the third law can also be made by considering the historical process by which Newton came to his third law. As Westfall (1971: 344–347) discusses, it is in Newton's attempt to solve the problem of collisions that he develops his concept of force. In this way we get (a) a measure of the external cause of changes of motion of a body, and (b) the separation of the concept of force from the concept of quantity of motion, and so from Descartes's law of conservation of motion for colliding bodies, giving us Newton's third law as the underpinning of the redistribution of the total quantity of motion (where quantity of motion is now the vector quantity momentum) in a collision, such that momentum is conserved.

15. Question (2) might take the form: in virtue of what is a body is hard (say), such that it can undergo collisions? The Cartesian reductionist seeks to reduce hardness to shape, size, and motion. In this way, hardness can be defined (in terms of shape, size, and motion) prior to its use in giving the laws of collision. However, it was already clear to many that the "strong program" is not going to work, and a weaker version of the project rejects the Cartesian restriction to shape, size, and motion and seeks to identify the appropriate properties to include in the reduction base. Once this move is made, question (2) becomes problematic: it is no longer clear what role an answer to this question has, even if one could be given.

16. It is a nonquestion except insofar as the law-constitutive approach can be brought to bear on it, but this style of tackling the question is not at all what Neile and his contemporaries had in mind: they were seeking a law-independent matter theory.

17. I am grateful to Daniel Garber for drawing my attention to this paragraph and to the article by Sophie Roux cited below.

18. Janiak (2008) argues that Newton rejects action-at-a-distance and that this remains unrevisable.

19. The material in this appendix was developed as part of a joint project with Dana Jalobeanu.

20. Again, there is a long history to this discussion in the *Principia* that can be found in Newton's manuscripts. For discussion of this history see Herivel 1965. Corollary IV reads: "The common center of gravity of two or more bodies does not alter its state of motion or rest by the actions of the bodies

among themselves; and therefore the common center of gravity of all bodies acting upon each other (excluding external actions and impediments) is either at rest, or moves uniformly in a right line)."
21. Or, as Newton writes: "Accordingly, as a result of equal changes in opposite directions in the motions of these bodies, and consequently as a result of the actions of the bodies on each other, the center is neither accelerated nor retarded nor does it undergo any change in its state of motion or of rest" (1999: 423).
22. Cohen (1980: 171–182) discusses this process in detail, and he attributes to the third law the role of allowing Newton to move from consideration of the motion of a single planet about a fixed center of force, to a pair of interacting planets, to a many-bodies interacting system, thereby constructing the motions of the planets in the manner we have described.

BIBLIOGRAPHY

Alexandrescu, V. and Jalobeanu, D., eds. (2003) *Esprits Modernes, Études sur les modèles de pensée alternatifs au XVI–XVIIe siècles*, Bucharest and Arad: University of Bucharest Press and Vasile Goldis University Press.
Brading, K. (forthcoming) "Newton's law-constitutive approach to bodies: a response to Descartes" in eds. A. Janiak and E. Schliesser forthcoming.
Broughton, J. and Carriero, J., eds. (2008) *A Companion to Descartes*, Oxford: Blackwell.
Cohen, I. B. (1980) *The Newtonian Revolution*, Cambridge: Cambridge University Press.
Descartes, R. (1991) *Principles of Philosophy*, trans. V. R. Miller and R. P. Miller, Dordrecht: Kluwer.
———. (1998) *The World and Other Writings*, trans. and ed. S. W. Gaukroger, Cambridge: Cambridge University Press.
Festa, E. and Gatto, R., eds. (2000) *Atomismo e continuo nel XVII secolo*, Naples: Vivarium.
Gabbey, A. (1980) "Force and inertia in the seventeenth century: Descartes and Newton" in ed. S. W. Gaukroger 1980, pp. 230–320.
Garber, D. (1992) *Descartes' Metaphysical Physics*, Chicago: University of Chicago Press.
Gaukroger, S. W., ed. (1980) *Descartes: Philosophy, Mathematics and Physics*, Brighton: Harvester Press.
———. (1995) *Descartes: An Intellectual Biography*, Oxford: Oxford University Press.
———. (2000) "The foundational role of hydrostatics and statics in Descartes' natural philosophy" in eds. S. Gaukroger, J. A. Schuster and J. Sutton 2000, pp. 60–80.
Gaukroger, S. W., Schuster, J. A., and Sutton, J., eds. (2000) *Descartes' Natural Philosophy*, London: Routledge.
Gueroult, M. (1980) "The metaphysics and physics of force in Descartes" in ed. S. W. Gaukroger 1980, pp. 196–229.
Herivel, J. (1965) *The Background to Newton's* Principia, Oxford: Oxford University Press.
Holden, T. (2004) *The Architecture of Matter: Galileo to Kant*, Oxford: Oxford University Press.
Jalobeanu, D. (2003) "The two cosmologies of René Descartes" in eds. V. Alexandrescu and D. Jalobeanu 2003, pp. 75–94.

Janiak, A. (2008) *Newton as Philosopher*, Cambridge: Cambridge University Press.

Janiak, A. and Schliesser, E., eds. (forthcoming) *Interpreting Newton: Critical Essays*, Cambridge: Cambridge University Press.

Lennon, T. M. (2007) "The Eleatic Descartes," *Journal of the History of Philosophy*, 45: 29–47.

Lennon, T. M. and Stainton, R. J., eds. (2008) *The Achilles of Rationalist Psychology*, Dordrecht: Springer.

McGuire, J. (1978) "Newton on place, time and God: an unpublished source," *British Journal for the History of Science*, 38: 114–129.

Newton, I. (1999) *The Principia: Mathematical Principles of Natural Philosophy*, eds. and trans. I. B. Cohen and A. Whitman, Berkeley: University of California Press; 1st edn 1687.

Normore, C. G. (2008) "Descartes and the metaphysics of extension" in eds. J. Broughton and J. Carriero 2008, pp. 271–287.

Radelet, P. (2000) "Axioms, principles and symmetries," manuscript.

Roux, S. (2000) "Descartes Atomiste?" in eds. E. Festa and R. Gatto 2000, pp. 211–274.

Rozemond, M. (2008) "The Achilles argument and the nature of matter in the Clarke Collins correspondence" in T. M. Lennon and R. J. Stainton 2008, pp. 159–175.

Westfall, R. S. (1971) *Force in Newton's Physics: The Science of Dynamics in the Seventeenth Century*, London: Macdonald.

8 Huygens, Wren, Wallis, and Newton on Rules of Impact and Reflection

Gemma Murray, William Harper, and Curtis Wilson

Newton cites, as evidence backing up the third law of motion,

> Law 3: *To any action there is always an opposite and equal reaction; in other words, the actions of two bodies upon each other are always equal and always opposite in direction.*
>
> (Newton 1999: 417)

the reports by Huygens, Wren, and Wallis, which were printed in the *Philosophical Transactions of the Royal Society* in 1668–1669. Huygens's and Wren's accounts deal with perfectly elastic collisions, Wallis's deals with perfectly inelastic collisions. Huygens, Wren, and Wallis do not provide details of experiments or any account for deviations from the ideal. Newton provides details of experiments and an account of how to extend the ideal theory to nonideal cases. He makes the law empirical rather than *a priori*. Though the experiments are complex and inexact they afford empirical support for the equality of action and reaction in collisions.

OUTLINE

1. Huygens's rules for the motion of bodies arising from mutual impact (perfectly elastic collision)
2. Wren's treatment of perfectly elastic collisions
3. Wallis's account of perfectly inelastic collisions
4. Newton's account of pendulum experiments taking into account air resistance and imperfectly elastic collisions

1 HUYGENS'S RULES FOR THE MOTION OF BODIES ARISING FROM MUTUAL IMPACT (PERFECTLY ELASTIC COLLISION)

1.1 Preface to Huygens's Rules by the Editor of the Philosophical Transactions

A Summary Account
of the Laws of Motion, *communicated by* Mr. Christian
Hugens *in a Letter to the* R. Society, *and since printed*
in French *in the* Journal des Scavans *of March* 18,
1669. st. n.

Before these *Rules of Motion* be here deliver'd, 'tis necessary to preface something, whereby the worthy Author of them may receive what is unquestionably due to him, yet without derogating from others, with whom in substance he agreeth. But, forasmuch as this Subject is of that nature, that all Philosophy and generally all Learn'd men are therein concern'd, it will be most proper, to publish these *Rules*, as well as we did those of D. *Wallis* and D. *Wren* (Numb. 43) in the Language of the Learn'd, together with some Historical passages relating thereto: Which we now doe, as follows.

(Huygens 1669: 925)

Translated by Curtis Wilson.

As in recent months several of the members of the Royal Society in their public assembly have urged strenuously that that most weighty subject of the Rules of Motion—not previously considered among them, other matters intervening so that it could never be discussed—be at last subjected to a strict examination, it seemed indeed good to that most illustrious company[1] that those in the Society who had meditated on the natural character of the motions to be investigated be asked to produce their meditations and discoveries on this subject, and at the same time, that the things thought out on this subject by other distinguished men such as Galileo, Cartesius, Honorato Fabri, Joachim Jung, Petrus Borelli, and others, be collected and edited, with this end in view, that the opinions of all being collected and consulted, the theory deriving therefrom and best agreeing with all observations and experiments (the latter done with the greatest care and good faith, and often repeated) be presented as is right to the Philosophic City.

This request being quickly promulgated, the first among the members of the learned society to be roused up were Christiaan Huygens, John Wallis, and Christopher Wren; they were spurred to mature and satisfactorily complete the hypotheses and rules of motion, in the composing of which they had long labored. So it was brought about that that select trio of outstanding men, in the space of only a few weeks, all but eagerly transmitted their theories, elegantly composed, and sought the opinion of the Royal Society concerning them.

First of all Dr. Wallis communicated for appraisal his Principles of Motion in a letter of 15 November 1668, and on November 29 lectured on them and handed them over. Soon Dr. Christopher Wren followed him, undertaking publicly to present to the Society his Law of Nature concerning the Collision of Bodies in the next month, on December 17. These two writings were then printed (with the prior consent of both authors), for the more convenient communication of these writings, and their wider discussion.

These matters having been attended to by us, lo! the letter-carrier brings to us on January 4 (Old Style) a letter from Dr. Huygens, and on January 5 (New Style), fully written out, his essay on the Motion of Bodies from Mutual Impulse, containing four Rules with their demonstrations.[2] Thus I had at hand the Wrennian account of the Theory, and immediately on the very same day, the public letter carrier helping, remitted the same, by way of compensation, to Dr. Huygens; delaying meanwhile the assignment [to the press] of the Huygenian writing (to include which, given its importance, and prior to the author's permission, would make me suspect), until the occasion arose for the most noble and wise president of the Royal Society, Vice-Comitem Brouncker, to order it. Which things being done, and the Rules of both being collated in the mode stated by the Society, immediately there appeared a marvelous consensus in both; which gave us the great pleasure of sending both writings to press. Nothing was wanting on the part of Huygens except his consent; without which, we judged, his writing must never be published, especially since he had not at that time given us the whole of it. It was meanwhile our care to insert his writing in the public memoirs of the Royal Society; at the same time on January 11 again thanking the author for his wise communication; then adding thereto, on February 4, a renewed request that he undertake to have his Theory printed either in Paris (most easily done in the Diaria Eruditorum, as they call it), or here in London in the Adversarii Philosophici; or at least permit it to be printed. Which being done, we received a little afterwards a second letter from Huygens, mentioning his receipt of Wren's piece on this subject, but saying nothing of preparing the publication of his own writing, either in Paris or in London.

Hence, I would say, it is altogether clear that Huygens himself failed in bringing about that publication, if not indeed giving occasion to procrastination. Therefore to the praised Dr. Wren, who in conformity with the sagacity of his genius brought forth a double Theory, the part in glory due to this Speculation properly should come. For beyond all doubt, neither of these [men] found out anything of these Theories from the other, before their writings were compared with each other; but each, by the fecundity of his own genius, brought forth this beautiful offspring.

Huygens indeed explained, when he was in London some years ago, those cases of motion which were then proposed to him; by a brilliant argument, in fact, he had found out Rules, the evidence for which he clearly presented. But he will not affirm that he has explained any part of his Theory to any of the English, unless it is claimed that he revealed it, being solicited by some of them to the communication of it; nor that ever, unless quite recently, he has caused it to be treated.

These things being offered with truth and justice, we thus present Huygens's Rules in the Latin language, for the more ample use of the learned.

1.2 Huygens's Rules for the Motion of Bodies Arising from Mutual Impact

Translated by Curtis Wilson.

1. If a hard body at rest is hit by an equal hard body in motion, after impact the impelling body will be at rest, and the velocity that was in the impelling body will have been acquired by the body originally at rest.
2. But if the first of the two equal bodies is also moved, and is borne in the same straight line, after contact the two of them will be moved with the velocities exchanged.
3. A body however great will be moved if impacted by any other body however small, with any velocity whatever.
4. The general rule for determining the motion that hard bodies acquire through direct impact is this:
    Let there be bodies A and B, of which A moves with velocity AD, and B is either moving in the same straight lines with velocity BD, or is at rest in B. Divide the line AB in C, the center of gravity of A and B, and let CE be taken equal to CD. I say that body A after impact will have velocity EA, and body B the velocity EB, each in the direction indicated by the order of the points EA, EB. Thus if E falls in the point A or B, body A or B will be brought to rest.
5. The quantity of motion of two bodies can increase or decrease through impact; but always there remains the same quantity in the same direction, when the quantity of the contrary motions is subtracted.
6. The sum of the products of the mass [*moles*] of each hard body multiplied into the square of its velocity is always the same before and after impact.
7. A hard body at rest accepts more motion from another hard body, itself greater or less than the body at rest, through the interposition of a third

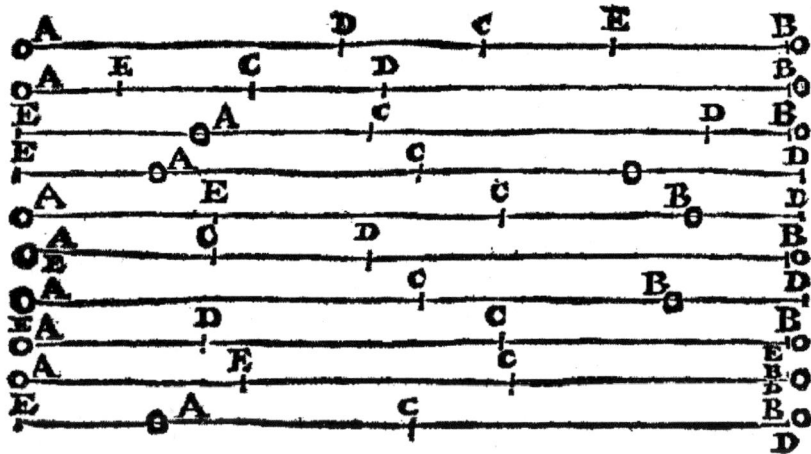

Figure 8.1 Huygens's diagram examples.

body which is intermediate in quantity, than if the percussion is immediate. And if the body interposed be a mean proportional between the other two, it acts more strongly on the body at rest.

[In a smaller type font, the editor adds:] In all these cases the Author considers bodies of the same material, or would have us estimate the mass [*moles*] from the weight.

He adds, moreover, that he has observed a certain wonderful law of nature, which he affirms he can demonstrate in spherical bodies, whatever velocity v be given, and in all others whether hard or soft, and whether impacting directly or obliquely, namely: The common center of gravity of two, three, or any number of bodies, always advances uniformly in the same direction and in a straight line, both before and after impact.

1.3 Comments on Huygens's Summary

Huygens's Rule 4:

4. The general rule for determining the motion that hard bodies acquire through direct impact is this: Let there be bodies A and B, of which A moves with velocity AD, and B is either moving in the same straight line with velocity BD, or is at rest in B. Divide the line AB in C, the center of gravity of A and B, and let CE be taken

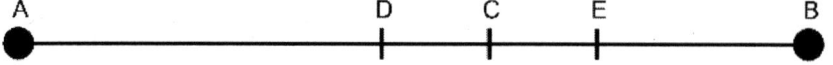

Figure 8.2 Huygens's basic diagram.

equal to *CD*. I say that body *A* after impact will have velocity *EA*, and body *B* the velocity *EB*, each in the direction indicated by the order of the points *EA*, *EB*. Thus if *E* falls in the point *A* or *B*, body *A* or *B* will be brought to rest.

This rule is a geometrical method of using line segments to give velocities after impact from velocities at impact, together with the center of gravity. The following diagram (Figure 8.3) is an illustration of Rule 4 and is based upon similar diagrams provided in the original publication.

The positions of the points A and B are fixed and the points D and C are placed somewhere on the line between them (D may also be placed on a continuation of this line to the left of A or to the right of B), depending on the initial velocities and masses of bodies A and B. The lengths of the lines AD and BD are proportional to the initial velocities of the bodies A and B. The ratio of the length AC to the length BC is equal to the ratio of the mass of body B to the mass of body A. Thus the lengths AC and BC represent the relative distance from each body to the center of mass of the composite system. This is the information required by Huygens's Rule 4 in order to calculate the velocities of the two bodies after collision, represented by the length and direction of the lines EA and EB. This is achieved using Huygens's diagram by using the lines AD, BD, AC, and BC to calculate the length and direction of the line CD, as shown in the diagram below. This line is, according to Rule 4, of equal length to CE. Once the position of E is found the final velocities EA and EB may be determined.

An example is given in Figure 8.4. The diagram represents a body A traveling toward a body B at 9 m/s and a body B traveling toward a body A at 11 m/s where body A has two thirds the mass of body B. The point D on Huygens's diagram can be calculated from converting these velocities into distances with a one-to-one ratio (1 m/s into 1m). Point C on Huygens's diagram can then be determined as representing the center of mass of the system if the distance between the bodies were as it is represented in the diagram. Thus AC can be calculated to be 12 m and BC 8 m. Using Huygens's diagram we can calculate the final velocities of both bodies by calculating the length CD by subtracting the length AD from the length AC, or equivalently by subtracting the length BC from the length BD. We can then locate point E on the diagram by use of the equivalence of the lengths of CD and CE. Once E is located we have the length and direction of the lines EA and EB, which are in a one-to-one ratio with the final velocities that would result from such a collision if it

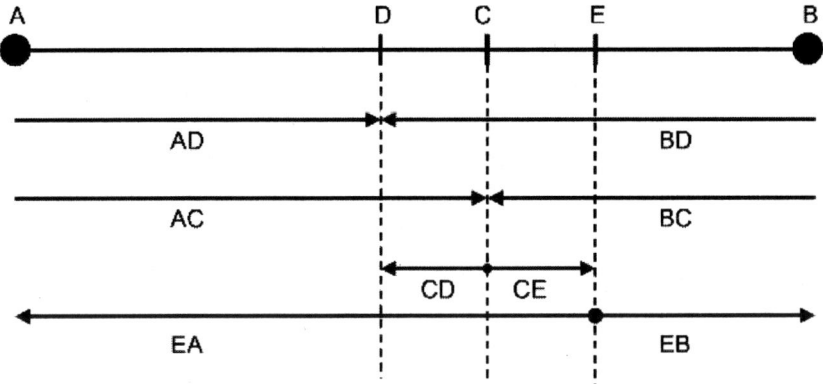

Figure 8.3 The basic construction.

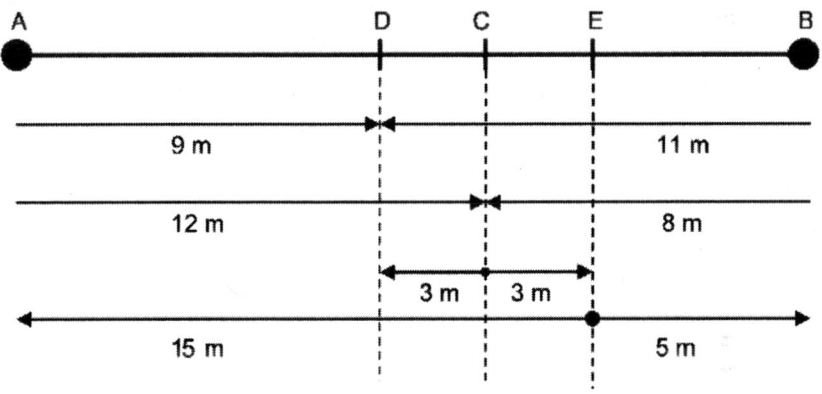

Figure 8.4 An example.

were perfectly elastic. Thus the final velocities of body A and body B can be found on Huygens's diagram as being 15 m/s and 5 m/s, both in the opposite direction to the original direction of motion.

Here are Huygens's next two rules. They give conservation of momentum and conservation of kinetic energy for such perfectly elastic collisions.

5. The quantity of motion of two bodies can increase or decrease through impact; but always there remains the same quantity in the same direction, when the quantity of the contrary motions is subtracted.

6. The sum of the products of the mass [*moles*] of each hard body multiplied into the square of its velocity is always the same before and after impact.

Let v stand for velocities before impact, and u for velocities after impact. This gives, for rule 5,

$$m_A v_A + m_B v_B = m_A u_A + m_B u_B$$

conservation of momentum, and for rule 6

$$m_A v_A^2 + m_B v_B^2 = m_A u_A^2 + m_B u_B^2,$$

conservation of kinetic energy.

Rule 4 follows from Rule 5 together with Rule 6. Indeed, given conservation of momentum for such collisions, the conservation of total velocity and the conservation of total kinetic energy are equivalent characterizations of such idealized perfectly elastic collisions. Rules 5 and 6 follow from Huygens's Rule 4.[3]

The following remarks, added by the editor in a smaller type font, are very interesting evidence that Huygens had, already by 1669, achieved, for collisions, a quite extraordinary understanding of some important fundamental concepts and results that Newton would present in his *Principia*.

> In all these cases the Author considers bodies of the same material, or would have us estimate the mass [*moles*] from the weight. He adds, moreover, that he has observed a certain wonderful law of nature, which he affirms he can demonstrate in spherical bodies, whatever velocity v be given, and in all others whether hard or soft, and whether impacting directly or obliquely, namely: The common center of gravity of two, three, or any number of bodies, always advances uniformly in the same direction and in a straight line, both before and after impact.

The first sentence suggests that Huygens anticipated Newton's distinction between mass and weight.[4] The second shows that, for collisions, Huygens anticipated corollary 4 of Newton's Laws of Motion.[5]

> **Corollary 4:** *The common center of gravity of two or more bodies does not change its state whether of motion or of rest as a result of the actions of the bodies upon one another; and therefore the common center of gravity of all bodies acting upon one another (excluding external actions and impediments) either is at rest or moves uniformly straight forward.*
>
> (Newton 1999: 421)

These remarks also indicate that Huygens understood that Rule 6 is restricted to "hard," i.e. elastic, bodies, while conservation of momentum applies in all cases of impact.

1.4 Huygens's Initial Incomplete Treatise

In the letter, which Oldenburg tells us he received and replied to on Jan 4 (Old Style), Huygens says that:

> What I have sent you is a beginning of a treatise on the motion of percussion, and the reason why I have chosen to begin with this topic in motion rather than with another is my desire to know the judgment of your illustrious members concerning my way of demonstrating, which however evident to me and to some of our members [in the Académie des Sciences] was unable to satisfy others . . .
> (Huygens 1888–1950, 6: 334–335)[6]

Huygens's essay contains detailed arguments that are not included in the above summary account that was printed in *Philosophical Transactions*.

HUYGENS'S HYPOTHESIS ON THE MOTION OF BODIES FROM MUTUAL IMPULSION

Translated by Curtis Wilson

1. Any body already in motion will continue to move perpetually with the same speed and in a straight line unless it is impeded.[7]
2. When two hard bodies, equal to each other and having equal speed, directly collide with one another, each rebounds with the same speed which it had before the collision.[8] By "directly colliding" it is meant that they move in the straight line joining their centers of gravity, and that the point of contact lies in this same straight line.
3. Both the motion of bodies and their equal or unequal speeds must be understood in relation to other bodies which are considered as being at rest, even if both sets of bodies happen to be involved in an additional common motion. As a result, when two bodies collide, then even if each of them is simultaneously subject to some other additional equal motion, they will act on each other quite as if the additional motion were absent.

For example, suppose a passenger in a ship, which is advancing with uniform motion, makes two equal balls collide with each other with equal speed as determined in relation to himself and to the parts of the ship. We say then that each ball must rebound with an equal speed in relation to the passenger carried along in the ship, as would clearly also happen if he were standing in the ship at rest or on the ground, and made the equal globes collide with equal speeds.

> 4. Whether I sustain the two bodies with my hands and make them collide with certain speeds, or another holding them gives to them the very same motions with respect to me, the rebounding of these bodies will be the same with respect to me.[9]

For instance, suppose I am standing still and holding with my hands A, B the bodies C, D suspended by strings; and suppose by moving my hands, I make the body C move with the speed CE, and the body D with the speed DE, so that they collide in E.

I say that the motions with which the two bodies rebound are the same with respect to me, whether the strings are held by my hands thus moving, or another person holds them with his hands and produces the same motions with respect to me.

With these things posited concerning the collision of equal bodies, we shall demonstrate by what laws they are mutually impelled. When we come to unequal bodies, we shall add some necessary hypotheses to what has already been said.

> **Proposition 1:** If a body collides directly with another equal body which is at rest, then after the contact the former is at rest, and the latter acquires the speed which was in the body that struck it.

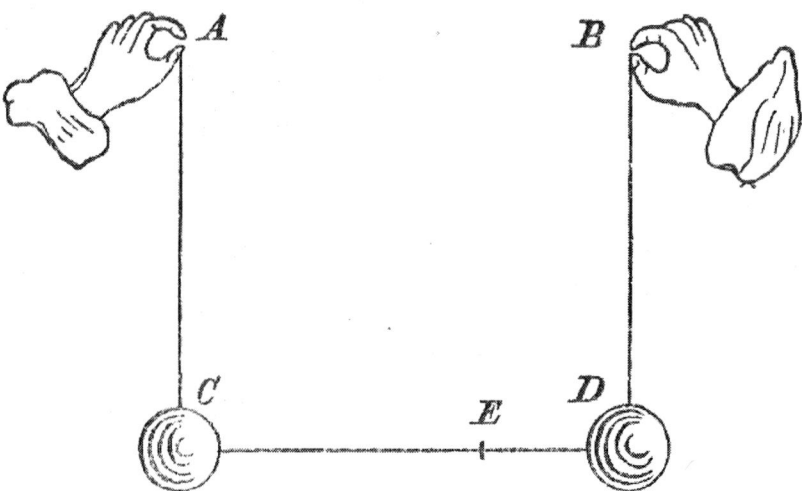

Figure 8a Huygens 1888–1950, 6: 337.

Rules of Impact and Reflection   163

Let there be the equal bodies E, F, suspended by the strings EH, FK, the ends of which strings someone holds with his hands; the hand H holds the body E unmoved, and the other hand K, together with the body F, is made to approach H so that the body F collides with E. I say that after impact the body F remains without motion, while E moves with the same speed previously belonging to F.

For let it be understood that he who holds up these bodies stands on the bank of the river, and let us imagine that a ship moves downstream close to the bank. Now let a passenger standing in the ship hold with his hands L, M the same bodies E, F which previously the experimenter on the riverbank held suspended by strings. Let the distance EF be divided in half at G, and suppose that the ship is moving to the left with the speed GE. Let the passenger cause the two bodies E, F to approach with equal speeds with respect to himself and the ship, so that they collide. They will then rebound with equal speeds necessarily, with respect to the passenger and the ship, by Hypothesis 3.

And since the hand L of the passenger is moved to the right with the same speed with which the ship moves to the left, it is clear that it

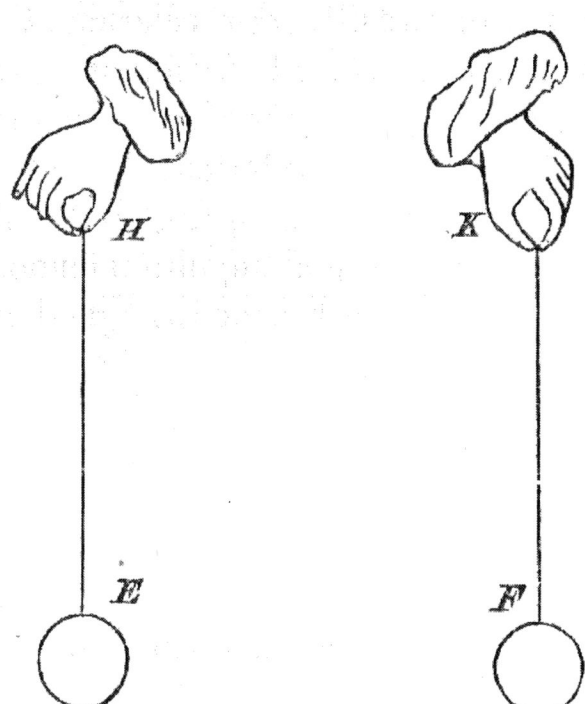

Figure 8b   Huygens 1888–1950, 6: 337.

will remain unmoved with respect to the riverbank and he who stands upon it, while the hand M, with respect to riverbank and the one standing there, will be moved with the speed FE, which is the double of GE...

And thus what happens here with respect to the reboundings of the balls E, F with respect to the one standing on the riverbank, would happen also if he held the ends of the strings, and held ball E immobile, and moved F with the speed FE, by Hypothesis 4.

For as we said, the balls E, F after contact rebound with equal speeds with respect to the passenger and the ship, ball E with speed GE, ball F with the speed GF, and meanwhile the ship moves toward the left with the speed GE. It follows that, with respect to the riverbank and the man standing on it, the ball F after impact remains without motion, and the ball E with respect to the same moves to the left with double of the speed GE, i.e. with the speed FE. Thus it is evident to the one standing on the riverbank that the same reboundings occur, so that the ball F after contact remains unmoved, and E advances with the speed FE, which previously F had. Q.E.D.

Proposition 2: If two equal bodies moved with unequal velocities collide, after contact they are moved with the speeds interchanged.

Let body E be carried with the speed EH to the right, and body F, equal to it, be carried with the lesser speed FH to the left. They will therefore

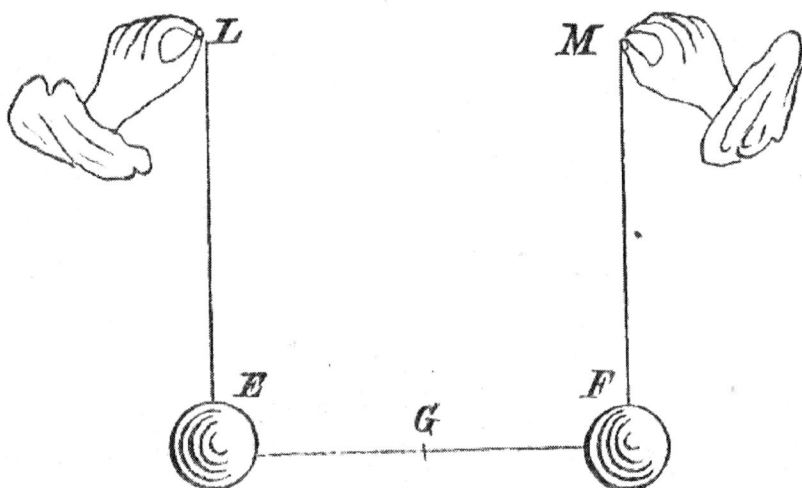

Figure 8c Huygens 1888–1950, 6: 338.

Rules of Impact and Reflection 165

meet in H, and after impact I say that body E will move with speed FH to the left, and body F with speed EH to the right.

Let the man standing on the riverbank effect the motions of the said bodies, holding with his hands C, D the ends of the strings by which they are suspended, and causing the hands, together with the bodies E, F, to come together with the speeds EH, FH. Let the distance EF be cut in half at G, and let it be understood that a ship is moving to the right with the speed GH. For a man standing in the ship, the ball E moves with the speed EG only, and the ball F with the speed FG; so that with respect to him the two balls approach their collision with equal speeds. On which account if it is supposed that the man standing in the ship grasps with his hands A, B the hands C, D of his associate standing on the riverbank, and with them the ends of the strings by which the balls are suspended, it will come about that he who stands on the riverbank will cause the bodies to collide with the equal speeds EG, FG [as measured by the man standing in the ship].

And thus with respect to the man in the ship, both balls will rebound from impact with equal speeds, E with speed GE, and F with speed GF. And the ship will meanwhile advance with the speed GH. And thus with respect to the riverbank and the man standing there, F will have the speed composed from both GF and GH, i.e., EH. And E will have the speed HF, the difference of the speeds GE, GH.

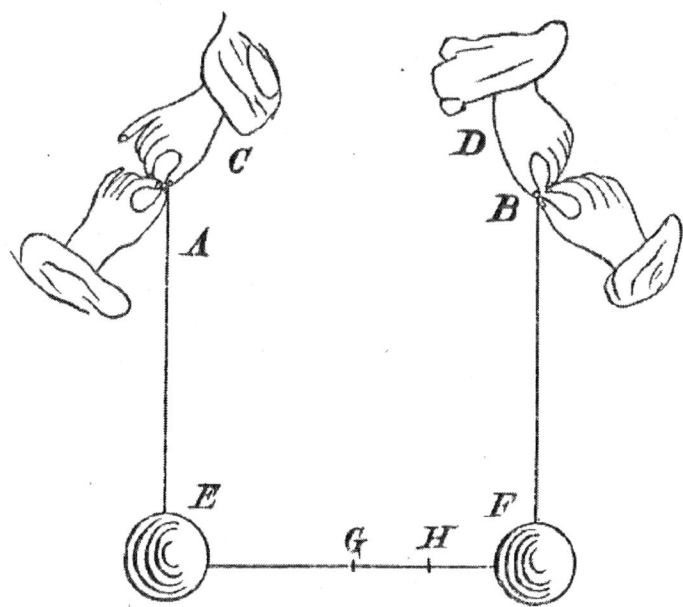

Figure 8d Huygens 1888–1950, 6: 339.

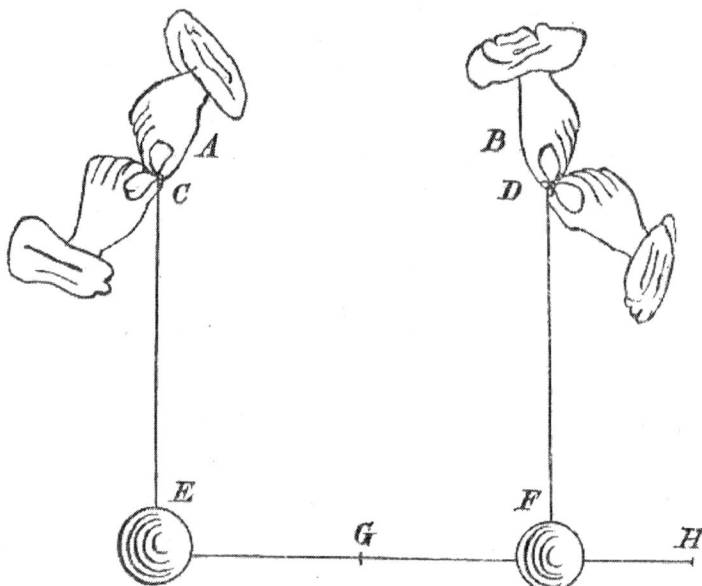

Figure 8e Huygens 1888–1950, 6: 340.

And thus we show that, for the man standing on the riverbank, the balls E and F impinging with the speeds EH, FH, will after impact have the speeds FH, EH respectively. Q.E.D.

If both bodies E and F move toward the right, E with the speed EH, and F in front with the smaller speed FH, then E follows F and they will meet in H. But I say that, after contact, F will have the speed EH, and E will follow with the speed FH. The demonstration is the same as in the preceding case.

Hypothesis 4: If a larger body meets a smaller body at rest, it gives some motion to the latter, and loses some of its own motion.

Proposition 3: A body however large when impacted by an arbitrarily small body with any speed, will be moved.

Suppose a ship to be moving alongside the riverbank, and in it a passenger stands and holds the bodies A and B suspended by strings. Let A on the left be the larger body, and B on the right the smaller. Let him hold B with his right hand D unmoved, with respect to both himself and the ship; meanwhile he moves rightward his left hand C toward D, and with it the body A, with any speed AB. Therefore B is impelled, and body A loses some of its motion, so that it proceeds toward the right with a smaller speed than AB. But while

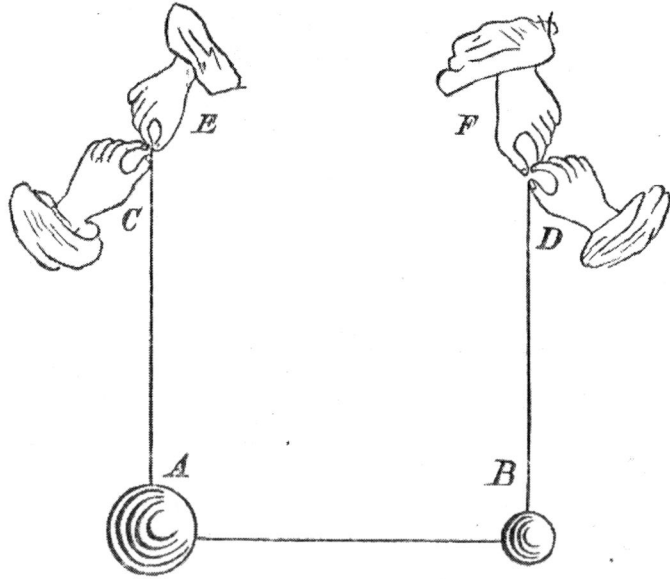

Figure 8f Huygens 1888–1950, 6: 341.

these things are occurring, it is supposed that the ship moves with the speed BA toward the left. Thus it comes about that as the passenger moves body A with the speed AB with respect to himself and the ship by which he is carried, the body A remains unmoved with respect to the riverbank and the spectator standing there, and equally the hand C. The other hand D with the body B will move with respect to the same spectator with the speed BA to the left, since with respect to the ship we supposed it to remain unmoved, and the ship is moved with the speed BA toward the left. Hence if the spectator standing on the riverbank is imagined to grasp with his hands E, F the hands of the passenger C, D, it will appear that while the ball A moves toward B, which is unmoved with respect to himself, simultaneously he will move B toward A, which is at rest with respect to himself and the riverbank. But we say that, from the impulse, the ball A, with respect to both the passenger and the ship, is borne to the right with a smaller speed than AB. But the ship is borne to the left with the speed BA. Therefore with respect to the riverbank and the spectator standing on it, it is manifest that, because of the impact, A is moved a little toward the left. And thus it is shown that a body A, however large, being taken as at rest by one standing on the riverbank, if impinged upon by a body B, however small, with any speed BA, will be caused to move. Q.E.D.

Proposition 4: Given two unequal bodies of which either one is in motion or both, and which collide head-on, and given the speeds of both, or of the one in motion if the other is at rest, to find the speeds with which they are moved after impact.

Let body A move to the right with speed AD, and let body B move in the opposite direction, or in the same direction, with the speed BD, or let it be at rest, so that point B falls on D. Thus the relative speeds of the two bodies will be AB.

Let AB be divided in C so that AC : CB :: magnitude B : magnitude A, and take CD equal to CE. I say that EA is the speed of A after impact, and EB the speed of body B, and this in the direction indicated by the order of the letters E A, E B. If point E falls on A, the body A is reduced to rest. If point E falls on B, the body B is at rest. For if we show that these things occur in a ship moving forward with uniform speed, it is clear that it will also take place for one who is standing on the shore. And thus let the ship be moved with respect to the riverbank, and let a passenger standing in the ship hold in his hands F, G the balls A, B suspended from strings, and moving these with the speeds AD, BD relative to himself and the ship, cause A, B to collide in D. But suppose the ship is moving with the speed DC in the direction indicated by the order of the letters D C. Thus it will happen that with respect to the riverbank and the spectator standing on it, the ball A will move with the speed AC toward the right, because with respect to the ship it will have the speed AD; but the ball B [Huygens has D!], since in the ship it has the speed BD, will have with respect to the riverbank the speed BC to the left. Therefore if a spectator standing on the riverbank grasps with his hands H, K the hands of the passenger F, G, together with the ends of the strings by which the bodies A, B are sustained, it is apparent that while the passenger moves them with respect to himself with the speeds AD, BD, at the same time he who stands on the riverbank will move them with respect to himself and the riverbank with the speeds AC, BC. Since these speeds are in the reciprocal ratio of the magnitudes [of the two bodies], it is necessary that the bodies A, B, with respect to the same spectator, rebound from contact with the same speeds CA, CB; this will be demonstrated later.[10] But the ship always advances with the speed DC or CE, according to the order of the points C E. Therefore it is necessary that A move, with respect to the ship and passenger, with the speed EA in the direction designated by the order of the points E A; B with respect to the same ship with the speed EB, likewise according to the order of the points E B.

But when E falls on A or on B, it is clear that body A or B after impact is moved with a speed equal to that of the ship itself, and in the same direction; whence in these cases it is necessary that it be at rest with respect to the ship and the passenger. And thus we show that the bodies A and B, which before impact moved in the ship with the speeds AD, BD, after impact move in the ship with the speeds E A, E B, according

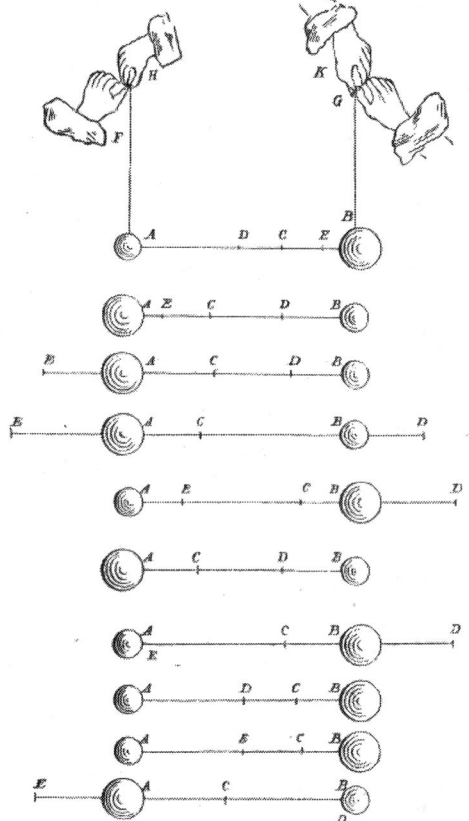

Figure 8g   Huygens 1888–1950, 6: 342.

to the order of these points. But what happens in the ship must certainly also happen for one standing on the shore. Therefore the proposition is clear.

1.5 Comments on Huygens's Initial Treatise

The above quoted passage from Huygens's letter points out that his thought experiment did not convince all his colleagues in the Académie des Sciences.[11] The paper that Huygens sent the Royal Society at Oldenburg's request in January 1669 was discussed by members of the Royal Society, who agreed that Huygens's rules coincided with Wren's.

This document is printed in the *Oeuvres Complètes* of Christiaan Huygens, 6: 334–344. It differs from the document translated from the French and published by Oldenburg in the *Philosophical Transactions* in April 1669 in a number of respects. It is in Latin. It begins with three premises (not labeled as Axioms or Hypotheses). A fourth paragraph then explains how the bodies may be suspended on strings, and held by a passenger on a ship, and observed by someone standing on the shore. If the ship is in motion with respect to the shore, the motions that the ship passenger causes the bodies to undergo are seen by the observer on shore as different motions with respect to himself, since the motion of the ship is superimposed on that of the bodies as determined by the ship passenger.

Propositions 1 and 2 then draw out the consequences of Premise 2 as implied by Premise 3. All the cases of equal bodies colliding with unequal speeds can be derived from comparisons between the motions of the bodies observed on shore compared with those observed by the ship passenger.

Next, Huygens introduces Hypothesis 4: If an arbitrarily large body collides with a smaller body that is at rest, it gives some motion to it, and thus loses some of its own motion. Proposition 3 then asserts that: A body however large, when impacted by a body arbitrarily small and moving with an arbitrary speed, is moved. The proof makes use again of the ship-and-shore observers. This proposition (and the corresponding Rule 3 of the summary published in *Philosophical Transactions*) appears to contradict Descartes's 4th Rule of impact.[12]

Proposition 4: Given two unequal bodies which collide directly with given speeds [one of them may be at rest], to find the velocities with which they rebound. Here Huygens gives all the rules that Wren gives. But like Wren, he *does not* derive his results from the Hypotheses so far stated.

The paper that Huygens published in the *Journal des Scavans* in March, and that Oldenburg translated into Latin and published in the *Philosophical Transactions* in April, includes a further hypothesis with the aid of which Proposition 4, or Wren's or Huygens's rules for the impact of unequal bodies, can be derived. Huygens informs us (Rule 6 of the cited document) that "The sum of the products of the magnitude of each hard body, multiplied by the square of the velocity, is always the same before and after impact." This proposition, combined with Galileo's relation between the square of the velocity and the height to which the body rises, is used in Huygens's final derivation of the general rule for the impact of unequal bodies. (This is found in the posthumous essay on the percussion of bodies.)

2   WREN'S TREATMENT OF PERFECTLY ELASTIC COLLISIONS

Dr. Christopher Wrens Theory concerning the same Subject; imparted to the R. Society Decemb.17 last, though entertain'd by the

Rules of Impact and Reflection 171

Author divers years ago, and verified by many Experiments, made by Himself and that other excellent Mathematician M. Rook before the said Society, as is attested by many Worthy Members of that Illustrious Body.

(Wren 1668: 867)

2.1 The Law of Nature Concerning the Collision of Bodies

Translated by Curtis Wilson.

The velocities of bodies that are proper and most natural are reciprocally proportional to the bodies.

And thus the bodies R, S having proper velocities retain proper velocities after impact.

And bodies R, S having improper velocities are restored by impact to equilibrium; that is, as much as R exceeds, and S falls short of the proper velocity before impact, by so much, as a result of the impact, is [the velocity of] R reduced and [that of] S increased, and *vice-versa*.[13]

Hence the collision of bodies having proper velocities is equivalent to a balance oscillating about its center of gravity.

And the collision of bodies having improper velocities is equivalent to a balance on two centers equally distant to this side and that from the center of gravity. The Yoke of the balance, where this is necessary, is produced.

And thus of equal bodies moving improperly there are three cases. Of unequal bodies improperly moving (whether in contrary directions or the same direction) there are altogether ten cases, of which five arise from conversion.[14]

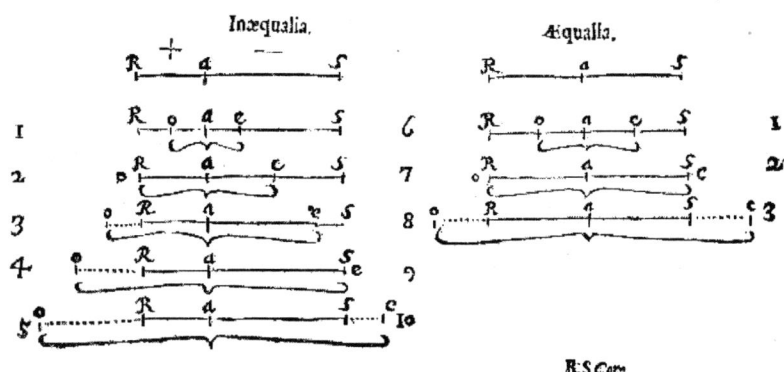

Figure 8.5 Wren 1668: 867.

R, S are equal bodies, or R is the greater body, S the lesser. The center of gravity or handle of the balance is a. Z is the sum of the velocities of the two bodies.

And, Re, Se = given velocities of the bodies before impact, and
  oR, oS = the sought velocities after impact.
Or, So, Ro = given velocities of the bodies before impact;
  eS, eR = the sought velocities after impact.
Calculus: R + S : S :: z : Ra, R + S : R :: Z : Sa.
  Re − 2Ra = oR, 2Sa ± Se = oS.
  So − 2Sa = eS, 2Ra + Ro = eR.

Nature observes the rules of addition and subtraction of Species [i.e. algebraic addition and subtraction].

2.2 Comments on Wren

Wren's rules make use of the terms "proper" and "improper." The speeds of two bodies about to collide are proper if they are inversely as the bodies, otherwise improper. If they are proper, the bodies after impact have speeds of the same size but reversed directions. If they are improper, the excess over the proper speed in the one body is transferred to the other body. This transfer does not render the speeds proper, but brings it about that the difference of speeds before impact is equal to the difference after impact, except for reversal of signs. Thus if the speeds before impact are v_1, v_2, and the speeds after impact are u_1, u_2, then

$$v_1 - v_2 = u_2 - u_1$$

The coefficient of restitution (as later defined by Newton) is thus 1. Perhaps the experimental results were close enough to this result that discrepancies from perfect restitution were not noticed?

The report on Wren gives his rules for elastic impact, and describes them as verified in many experiments. Wren cites earlier experiments he and Rooke had made before members of the Royal Society.[15] Hall describes some of these experiments, made during Huygens's visit to England in 1661, as ones that agreed with solutions calculated by Huygens.[16]

As we did for Huygens, we write today's equations for conservation of momentum and *force vive*:

1) $m_1 v_1 + m_2 v_2 = m_1 u_1 + m_2 u_2$,
2) $m_1 v_1^2 + m_2 v_2^2 = m_1 u_1^2 + m_2 u_2^2$.

From these two rules we can deduce that $v_1 - v_2 = u_2 - u_1$, or the velocity lost by m_1 is equal to the velocity gained by m_2, as Wren claims.

Rules of Impact and Reflection   173

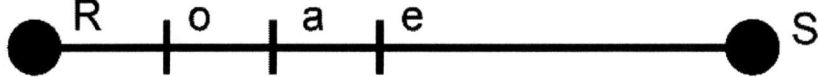

Figure 8.6   Wren's basic diagram.

Wren's diagrams and his use of them to determine the resultant velocities of two bodies subject to a perfectly elastic collision are remarkably similar to Huygens's.

A notable addition in Wren's account is his discussion of the symmetries present in the system. He notes that Re and Se may be chosen to represent the initial velocities of the bodies, and oR and oS the final velocities, or equivalently Ro and So may represent the initial velocities and eR and eS the final velocities. This equivalence is illustrated in Figures 8.7 and 8.8, where in both figures the top line represents the initial velocities, the second, the bodies' distances from their joint center of mass, the third the equivalence between the distance of the line oa and the line ae, and the bottom line the final velocities of the two bodies after the collision.

Wren also provides us with some sort of categorization of the possible two-body elastic collisions in his eight diagrams, five of bodies of unequal mass and three of equal mass. These diagrams make clearer the role of the equality of the lengths of oa and ae in determining the final velocities.

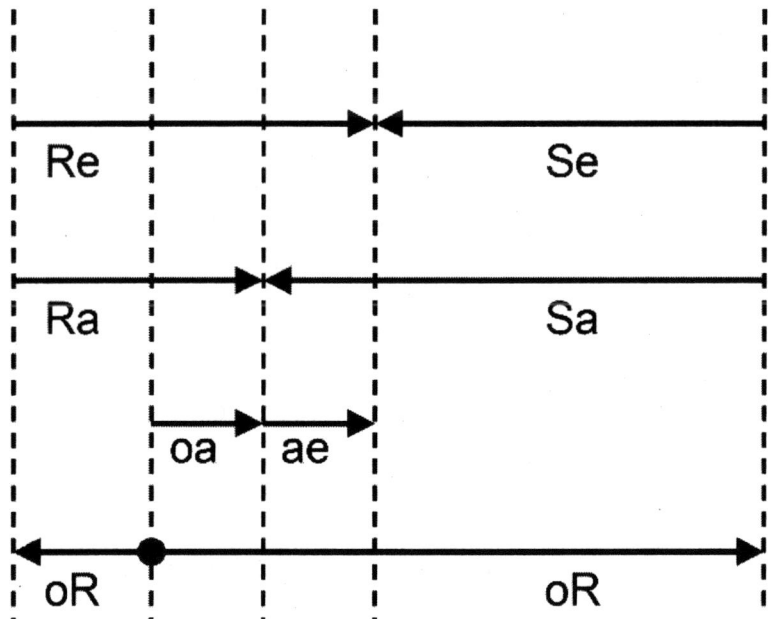

Figure 8.7   Re, Se initial velocities; oR's final velocities.

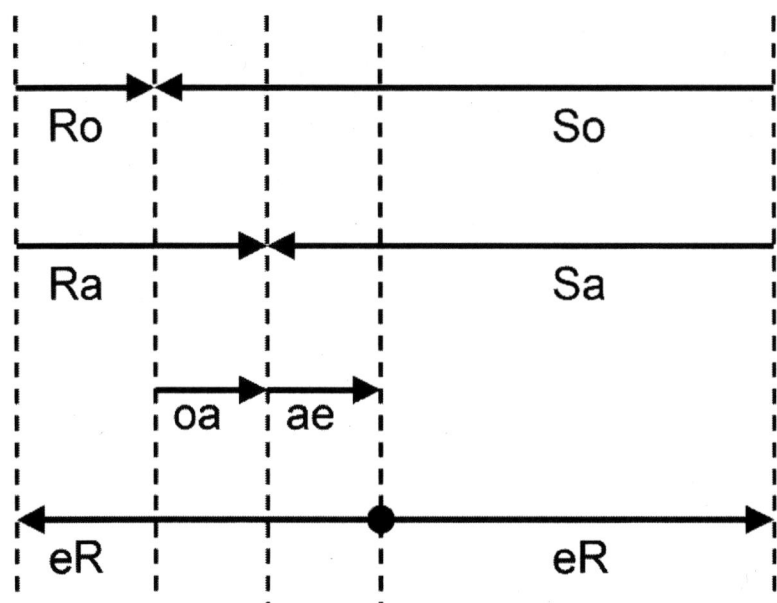

Figure 8.8 Ro, So as initial velocities; eR's as final velocities.

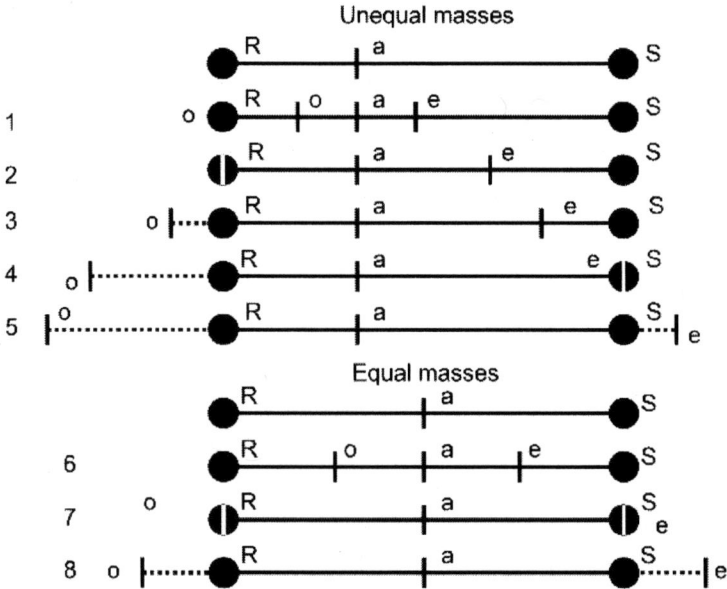

Figure 8.9 Wren's diagram examples exploiting symmetries.

3 WALLIS'S ACCOUNT OF PERFECTLY INELASTIC COLLISIONS

A Summary account by Dr. John Wallis, of the General Laws of Motion, by way of Letter written by him to the Publisher, and communicated to the R. Society, Novemb. 26, 1668.

(Wallis 1668: 864)

3.1 Wallis's Text

Translated by Curtis Wilson.

You ask, my dear Sir, that in a few words I reveal what are my principles for evaluating motions. This same thing, if you remember, was already done previously, not only in that Opus that was exhibited at the Royal Society eight months ago, and was by their command printed; but also long ago in two writings exhibited at the same society several years ago, and which you also have in your possession. One of them explains, from the general principles of motion, how it is possible for a man by his own breath, inflating a bladder, to raise at least 100 pounds; which experiment was demonstrated 16 or 18 years ago at Oxford, and there several times repeated. The other expounds various experiments of those called Torricellian, from hydrostatic principles.

The sum of the matter reduces to this.

1. If an Agent A has an effect E, an Agent 2A will have an effect 2E, an Agent 3A an effect 3E, and so on, other things being equal. Universally, mA will be as mE, whatever the coefficient m.
2. Therefore if a force V moves a weight P, a force mV will move mP, other things being equal; for instance, through the same length in the same time; in other words, with the same velocity.
3. For instance, if in time T it moves the weight through length L, in time nT it will move the weight through nL.
4. And thus if force V in time T moves weight P through length L; the force mV in time nT will move mP through length nL. And therefore as VT (the product of the forces and time) is to PL (the product of weight and length), so is mnVT to mnPL.
5. Since the degrees of velocity are proportional to the lengths traversed in the same time, or (and this comes to the same thing) reciprocally proportional to the time expended in traversing the same length, therefore

$$\frac{L}{T} : C :: \frac{mL}{nT} : \frac{m}{n}C.$$

That is, the degrees of velocity are in the ratio compounded from the lengths directly and from the times reciprocally.

6. Therefore, since VT : PL :: mnVT : mnPL, we shall have

$$V:\frac{PL}{T} :: mV:\frac{mnPL}{nT};$$

that is, V : PC :: mV : mPC, where mP × C = P × mC.

7. This means that, if force V is able to move weight P with velocity C, force mV will either move the same weight P with the velocity mC, or the weight mP with the same velocity C; or in sum, whatever weight with that velocity such that the product of weight and velocity is mPC.

8. Thence depends the ratio for constructing all machines for facilitating motion: in whatever ratio the weight is increased, in the same ratio the velocity is decreased. By which it comes about that the effect is as the product of weight and velocity, when the same force is the mover; e.g.

$$V : PC :: V : mP \times \frac{1}{m}C.$$

9. If a weight P, borne by a force V with velocity C impinges directly on a weight mP at rest (but not impeded), each is borne forward with the velocity

$$\frac{1}{1+m}C.$$

For, with the same force applied to moving a greater weight, the velocity is decreased in the ratio of the increase [in weight]. Thus

$$V : PC :: V : \frac{1+m}{1}P \times \frac{1}{1+m}C.$$

And thus the one impetus (arising from weight and velocity) will be

$$\frac{1}{1+m}PC;$$

the other will be

$$\frac{1}{1+m}mPC.$$

10. If on a weight P, borne by force V with a velocity C, there impinges directly another weight, along the same path, following with a greater velocity; e.g. weight mP with velocity nC (therefore borne by force mnV), each of them will be borne with the velocity

$$\frac{1+mn}{1+m}C.$$

For V : PC :: mnV : mnPC, and therefore [componendo] V : PC :: V + mnV : PC + mnPC :: (1 + mn)V : (1 + mn)PC. But

$$(1+mn)\text{PC} = \frac{1+m}{1}\text{P} \times \frac{1+mn}{1+m}\text{C}.$$

Therefore the impetus of the preceding body will be

$$\frac{1+mn}{1+m}\text{PC};$$

of the following body,

$$\frac{1+mn}{1+m}m\text{PC}.$$

[Translator's note: Wallis first obtains the velocity of the two bodies after impact: it must be such that the total weight P + mP has after impact a single velocity v such that the total impetus is equal to the sum of the original impetuses, or (P + mP)v = PC + mnPC. When v is known, so will be the impetus of each of the two bodies, since their weights are known.]

11. If weights borne by opposite forces directly impinge on one another, e.g. weight P moved by force V with velocity C toward the right, and weight mP with velocity nC (and therefore with force mnV) toward the left, the velocity, impetus, and direction of each of them are deduced as follows. The weight borne toward the right, if with the other it should come to be [relatively] at rest, would introduce the velocity, and therefore the impetus

$$\frac{1}{1+m}m\text{PC}$$

toward the right, and would itself retain this same velocity, and hence the impetus

$$\frac{1}{1+m}\text{PC}$$

toward the right (by section 9). The weight borne toward the left, by a similar reasoning, if it should come to be [relatively] at rest with the other, would introduce a velocity

$$\frac{mn}{1+m}\text{C},$$

and therefore the impetus

$$\frac{mn}{1+m}\text{PC},$$

toward the left; and would retain for itself this same velocity, and therefore the impetus

$$\frac{mn}{1+m}m\text{PC}$$

toward the left. And since each motion occurs thus; the earlier impetus to the right will now be aggregated from

$$\frac{1}{1+m}\text{PC}$$

to the right, and

$$\frac{mn}{1+m}\text{PC}$$

to the left; and therefore in itself either to the right or to the left, according as the former or the latter is greater, with an impetus which is the difference of the two. That is, with + signifying rightward, and − signifying leftward, the impetus will be

$$+\frac{1}{1+m}\text{PC} - \frac{mn}{1+m}\text{PC} = \frac{1-mn}{1+m}\text{PC}.$$

The velocity will be

$$\frac{1-mn}{1+m}C,$$

and therefore either to the right or to the left, according as 1 or mn is greater. And similarly the earlier impetus leftward will be

$$+\frac{1}{1+m}m\text{PC} - \frac{mn}{1+m}m\text{PC} = \frac{1-mn}{1+m}m\text{PC}.$$

The velocity will be

$$\frac{1-mn}{1+m}C,$$

and either to the right or the left, according as 1 or mn is greater.

12. If the weights neither proceed in the same direct way, nor directly opposite, but impinge on each other obliquely, the preceding calculation must be altered in the measure of the obliquity. The impetus of a body impinging obliquely, is to the impetus if the body impinged directly, other things being equal, as the radius to the secant of the angle of obliquity. This is to be understood as well where [the body] falls perpendicularly, but hits the surface obliquely, as when the

paths of the motions intersect obliquely. Which consideration, with the prior calculation applied as required, will determine the velocity, impetus, and direction of the obliquely impinging bodies; that is, with what impetus and what velocity, and in what parts [directions] the bodies which thus impinge will rebound. And the ratio is the same of the gravity of heavy bodies descending obliquely to that of the same bodies descending perpendicularly. As we have demonstrated elsewhere.

13. If the bodies that impinge are understood to be not absolutely hard (as we have previously supposed) but ceding to the blow so that by their elastic force they are able to restore themselves, it can happen that these bodies mutually rebound, which would otherwise proceed together, and indeed more or less, according as this restituting force is greater or less, namely if the impetus from the restituting force is greater than the progressive.

In accelerated and retarded motions, the impetus in particular moments is to be considered that which agrees with the degree of velocity then acquired. But where the motion is along a curve, its direction is to be taken at any particular moment as that of the straight line touching the curve where the body is. And if the motion is accelerated and retarded, and also curved, as in the vibration of pendulums, the impetus is to be taken, in single points, according to the degree of acceleration and the obliquity of the tangent. And these are, as I judge, the general laws of motion, which are to be accommodated to particular cases by calculation. Which cases, were I to pursue them in detail, I would transgress the limits of a letter. Nor can one easily know them with the aid of diagrams, from which I here thought I should abstain. Farewell. Oxford, 15 November 1668.

3.2 Comments on Wallis's Account

Wallis seems to have obtained his rules from the rules for simple machines in the Medieval Science of Weights.[17] Here the weights are inversely as the virtual velocities of the weights connected in the constraint system. Thus the "impetus" of a weight can be measured by the product of the weight P by its velocity C. There is equilibrium when the impetuses of two weights are equal:

$$P_1 C_1 = P_2 C_2$$

Can this schema be applied to inelastic impact? Wallis proceeds as if it applies. Suppose the two bodies before colliding have the impetuses $P_1 C_1, -P_2 C_2.$ Their total impetus before collision will be

$$P_1C_1 - P_2C_2$$

After impact, the total impetus will remain the same, the total impetus being shared between the two bodies in proportion to their weights, so that they are, respectively,

$$\frac{P_1}{P_1+P_2}(P_1C_1 - P_2C_2) \quad \frac{P_2}{P_1+P_2}(P_1C_1 - P_2C_2)^{18}$$

Wallis's account of the motion of bodies resulting from inelastic collisions is the result of a derivation from first principles involving the scaling of forces and their effect on the motions of bodies. This derivation leads to the key principle of Wallis's account, his Rule 8:

> 8. Thence depends the ratio for constructing all machines for facilitating motion: in whatever ratio the weight is increased, in the same ratio the velocity is decreased. By which it comes about that the effect is as the product of weight and velocity, when the same force is the mover; e.g.
>
> $$V : PC :: V : mP \times \frac{1}{m}C.$$

Rule 8 is a form of the law of conservation of momentum (which is called "impetus" by Wallis). Since Wallis's account only deals with the cases of collision in which the two bodies effectively become one joint-body with a mass equal to the sum of the masses of the initial two bodies, Rule 8 provides the only necessary principle for determining the velocity of the body after the collision. Since the two bodies become one joint-body we do not require a rule to determine how the momentum of the two bodies before the collision is distributed between the two bodies after the collision. The momentum of the two bodies before the collision is simply equal to the momentum of the joint-body after the collision. Thus, in order to calculate the final velocity of the joint-body we need only determine what velocity of that joint-body will be in accord with the law of conservation of momentum.

This Rule 8 is applied to three different types of two-body collisions by Wallis in his points 9, 10, and 11. The case described in Wallis's point 9 is one in which one body collides with another body that is at rest. This case is illustrated in figure 10.

In this example Body 1 has weight P and Body 2 has weight mP. The initial velocity of Body 1 is C and Body 2 is stationary. Wallis then calculates the resultant velocity of the two bodies together by applying his rule 8. The resultant velocity is found by calculating the momentum of the system before the collision and determining what final velocity will result in the same impetus after the collision.

Figures 8.11 and 8.12 illustrate Wallis's examples 10 and 11.

Wallis supposes his idealized bodies to be "absolutely hard." He tells us that his idealized hard bodies are "not such that they cede to the blow so that by their elastic force they are able to restore themselves." Elasticity requires some resilience in the parts of the body, "by which it can happen that these bodies mutually rebound." Deformation that is not restored requires loss of energy. Wallis's idealized bodies, apparently, are to remain rigid throughout the interaction. Wallis doesn't show any interest in empirically verifying these rules.

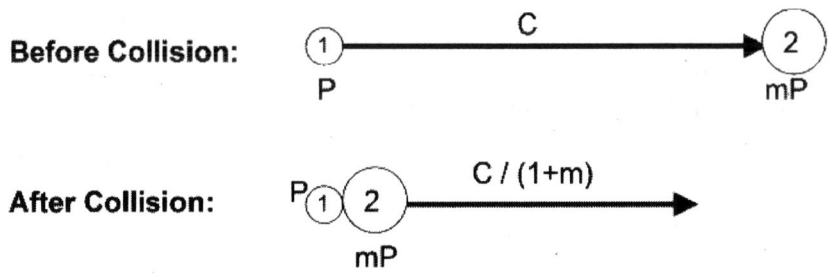

Figure 8.10 Body 1 hits stationary body 2.

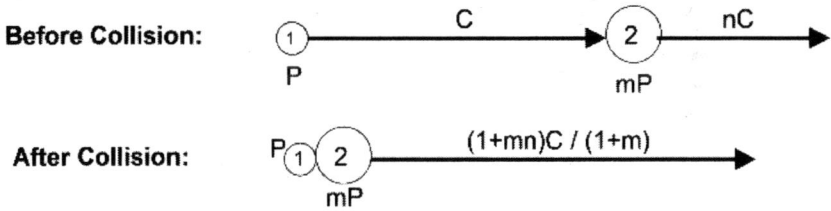

Figure 8.11 Body 1 overtaking body 2.

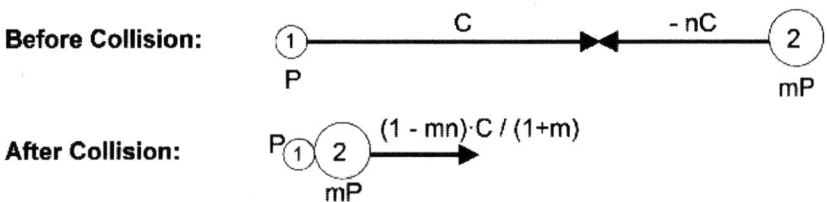

Figure 8.12 A head-on collision.

4 NEWTON'S ACCOUNT OF PENDULUM EXPERIMENTS TAKING INTO ACCOUNT AIR RESISTANCE AND IMPERFECTLY ELASTIC COLLISIONS

In his scholium to the Laws Newton, after citing support for the first two laws of motion and the first two corollaries from pendulums and clocks, appeals to the work in which Wren, Wallis, and Huygens found the rules for the collisions and reflections of hard bodies.

> From the same laws and corollaries and law three, Sir Christopher Wren, Dr. John Wallis, and Mr. Christiaan Huygens, easily the foremost geometers of the previous generation, independently found the rules for the collisions and reflections of hard bodies, and communicated them to the Royal Society at nearly the same time, entirely agreeing with one another (as to these rules); and Wallis was indeed the first to publish what had been found, followed by Wren and Huygens. But Wren additionally proved the truth of these rules before the Royal Society by means of an experiment with pendulums, which the eminent Mariotte[19] soon after thought worthy to be made into the subject of a whole book.
>
> (Newton 1999: 424–425)

The cited pendulum experiment would have provided measurements of the equality of actions and reactions for applications of law 3 to collisions.

Newton points out that such an experiment ought to take into account air resistance and the elastic force of the colliding bodies.

> However, if this experiment is to agree precisely with the theories account must be taken of both the resistance of the air and the elastic force of the colliding bodies. . . .
>
> (Newton 1999: 425)

The rest of the paragraph, which is just over a whole page in length, gives the detailed method by which Newton carried out his own improved version of this experiment to take into account air resistance.

He uses an example of a pendulum experiment in which a body A is let fall to collide with a stationary body B, to show how to correct for air resistance to find the arcs of motion that represent the velocities before and after collision that would have obtained if the experiment had been performed in a vacuum.

He shows how to calculate the arc TA that would, in a vacuum, have produced the velocity that body A let go from S actually has when it contacts body B. He also shows how to calculate the arcs At, and Bl that would have resulted after reflection had the actual experiment, which did result in A going to s and body B going to k after reflection, been performed in a

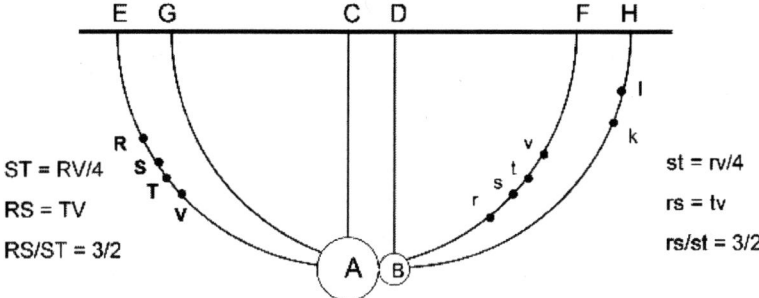

Figure 8.13 Newton's diagram with notes added.

vacuum. This method is described in detail with reference to the diagram above by Newton.

> Let the spherical bodies A and B be suspended from centers C and D from parallel and equal cords AC and BD. With these centers and those distances as radii describe semi-circles EAF and GBH bisected by radii CA and DB. Take away body B, and let body A be brought to any point R of the arc EAF and be let go from there, and let it return after one oscillation to point V. RV is the retardation arising from the resistance of the air. Let ST be a fourth of RV and be located in the middle so that RS and TV are equal and RS is to ST as 3 is to 2. Then ST will closely approximate the retardation in the decent from S to A.
>
> (Newton 1999: 425)

> Restore body B to its original place. Let body A fall from point S, and its velocity at the point of reflection A, without sensible error, will be as great as if it had fallen in a vacuum from place T.
>
> (Newton 1999: 425)

Newton determines the effect of air resistance on body A in moving through one whole oscillation by noting the distance between the point R, where body A begins its motion, and point V, where the body arrives upon completion of the oscillation. The arc RV represents the retarding effect of air resistance on the body in the completion of an oscillation. Newton then calculates the effect on the first part of this oscillation, which the body will move through before a collision at point A. He determines this by dividing the arc RV, the measured retardation for a whole oscillation, by four and averaging the length of the path during each of the four parts of the whole oscillation. This gives us the arc ST, which is one quarter of the length of RV and placed in the middle of

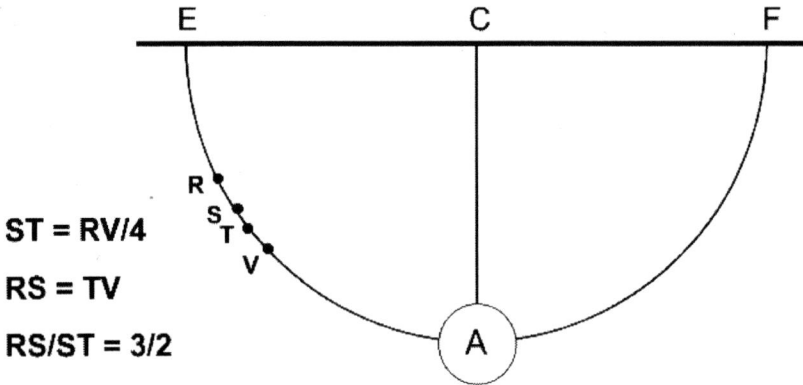

Figure 8.14 Diagram illustrating Newton's calculation.

this original arc. The arc ST represents the effect of air resistance on the body A completing the arc SA. Newton goes on to utilise these arcs to represent velocities.

> Therefore let this velocity be represented by the chord of the arc TA. For it is a proposition very well known to geometers that the velocity of a pendulum in its lowest point is as the chord of the arc that it has described in falling.[20]
>
> After reflection let body A arrive at place s, and body B at place k. Take away body B and find the place v such that if body A is let go from this place and after one oscillation returns to place r, st will be a fourth of rv and be located in the middle so that rs and tv are equal and let the chord of the arc tA represent the velocity that body A had in place A immediately after reflection. For t will be that true and correct place to which body A must have ascended if there had been no resistance of the air.
>
> By a similar method the place k, to which body B ascends, will have to be corrected, and the place l, to which that body must have ascended in a vacuum will have to be found.
>
> In this manner it is possible to make all our experiments, just as if we were in a vacuum. Finally body A will have to be multiplied (so to speak) by the chord of the arc TA, which represents its velocity, in order to get its motion in place A immediately before reflection, and then by the chord of the arc *t*A in order to get its motion in place A immediately after reflection. And thus body B will have to be multiplied by the chord of the arc B*l* in order to get its motion immediately after reflection. And by a similar method when two bodies are let go simultaneously from different places, the motions of both will have to be found before as well as after reflection, and then the motions

will have to be compared with each other in order to determine the effects of the reflection.

(Newton 1999: 425–426)

On the by then well-known, idealized theory of uninterrupted pendulum motion in a vacuum, the velocity of a pendulum in its lowest point is as the chord of the arc that it has described in falling and as the chord of the equal arc it traverses from that lowest point to its highest point on the other side. In the application to collisions, these chords of the arcs before and after reflection represent the velocities before and after that collision. There is no attempt to treat the details of the transition from the state of motion before to the state of motion after reflection.

These experiments give theory-mediated measurements of the equality of action and reaction corresponding to applications of the third law of motion to collisions.

> On making a test in this way with ten-foot pendulums, using unequal as well as equal bodies, and making the bodies come together from very large distances apart, say of eight or twelve or sixteen feet, I always found—within an error of less than three inches in the measurements—that when the bodies met each other directly, the changes of motions made in the bodies in opposite directions were equal, and consequently that the action and reaction were always equal. . . .
>
> (Newton 1999: 426)

Examples of a collision with one body striking another at rest and of two bodies colliding head-on are followed by the following description of a collision where one body overtakes another moving more slowly in the same direction:

> But if the bodies moved in the same direction, A more quickly with fourteen parts and B more slowly with five parts, and after reflection A moved with five parts, then B moved slowly with five parts, and after reflection A moved with five parts, then B moved with fourteen, nine parts having been transferred from A to B.
>
> (Newton 1999: 426)

Newton summarizes the agreement of all these empirical measurements with the equality of action and reaction:

> As a result of the meeting and collision of bodies, the quantity of motion—determined by adding the motions in the same direction and subtracting the motions in opposite directions—was never changed. I would attribute the error of an inch or two in the measurements to the difficulty of doing everything with sufficient accuracy.
>
> (Newton 1999: 426)

These experiments are theory-mediated measurements that show results that accurately agree with the equality of action and reaction in collisions, within the precision to which the velocities can be empirically determined. The theory of pendulum motion made it possible to carry out these velocity measurements.

Newton goes on to extend these pendulum experiments to bodies that are not perfectly elastic:

> Further, lest anyone object that the rule which this experiment was designed to prove presupposes that bodies are either absolutely hard or at least perfectly elastic and thus of a kind which do not occur naturally, I add that the experiments just described work equally well with soft bodies and with hard ones, since surely they do not in any way depend on the condition of hardness. For if this rule is to be tested in bodies that are not perfectly hard, it will only be necessary to decrease the reflection in a fixed proportion to the quantity of elastic force. In the theory of Wren and Huygens, absolutely hard bodies rebound from each other with the velocity with which they have collided. This will be affirmed with more certainty of perfectly elastic bodies. In imperfectly elastic bodies the velocity of rebounding must be decreased together with the elastic force, because that force (except when the parts of the bodies are damaged as a result of the collision, or experience some sort of extension as would be caused by a hammer blow) is fixed and determinate (as far as I can tell) and makes bodies rebound from each other with a relative velocity that is in a given ratio to the relative velocity with which they collide.
>
> I have tested this as follows with tightly wound bound balls of wool strongly compressed. First releasing the pendulums and measuring their reflection I found the quantity of their elastic force; then from this force I determined what the reflections would be in other cases of their collision, and the experiments which were made agreed with the computations. The balls always rebounded from each other with a relative velocity that was to the relative velocity of their colliding as 5 to 9, more or less. Steel balls rebounded with nearly the same velocity and cork balls with a slightly smaller velocity, while with glass balls the proportion was roughly 15 to 16.
>
> (Newton 1999: 427)

Ball Material	Velocity after collision / Velocity before collision
Steel	~1
Cork	Slightly less than 1
Glass	15/16
Tightly wound wool	5/9

These results indicate that, for the velocities obtained in Newton's pendulum experiments, there were no detectable losses corresponding to imperfect elasticity of the steel balls. For such velocities, the experiments count as measurements of a coefficient of restitution for steel that was not detectably different from 1. Newton's report of obtaining 15/16 for glass balls and slightly less than 1 for cork balls suggests that his experiments were accurate enough to have been able to detect fairly small deviations from perfect restitution had they occurred.

These experiments afford theory-mediated measurements of velocities that are fit by the equality of action and reaction calculated in accordance with the reductions appropriate to the differing elasticities. This suggests that the third law of motion applies to collisions quite generally. What are to be counted as the action and reaction are to be appropriately reduced to take into account losses due to imperfect elasticity, as well as any losses due to damage of the bodies in the collision.[21]

The passages describing Newton's pendulum experiments are given in his scholium to the laws of motion. Newton opens this scholium by describing his laws of motion as "accepted by mathematicians and confirmed by experiments of many kinds" (Newton 1999: 424). Newton clearly counts the experiments he has described as among the experiments of many kinds that he claims confirm his laws of motion.

5 CONCLUSION

Huygens takes laws of collision between bodies as fundamental contributions to his project of making motion phenomena intelligible by showing how they could be accounted for by contact action between bodies. These laws may be regarded as informing his conception of bodies. Huygens and Wallis put considerable weight on a priori reasoning. Wren and Newton explicitly appeal to experiments, though Wren does not provide experimental details. Like Huygens and Wallis, Wren restricts his account to idealized cases. Newton provides an account of experimental details and an account of how to use experiments to extend the ideal theory to nonideal cases, where he takes into account air resistance and imperfect elasticity. Though the experiments are complex and inexact they appear to afford significant empirical support for the equality of action and reaction in collisions.[22]

NOTES

1. Wilson and Harper thank Vicki Harper and Richard Olson for help with the translation of "caetui" as "company."
2. See section 1.4 below.
3. A derivation of conservation of momentum from Huygens's diagrams and his Rule 4:

From Huygens's diagram: $AC - AD = AE - AC \Rightarrow EA = AD - 2AC$
$BD - BC = BC - BE \Rightarrow EB = BD - 2BC$

The centre of mass C of A and B is given by the following equation where MA is the mass of body A and MB is the mass of body B: $MA \cdot AC + MB \cdot BC = 0$

From the equations from Huygens; diagrams:

$AC = (AD - EA)/2, BC = (BD - EB)/2$

These equations give us:

$MA \cdot (AD - EA)/2 + MB \cdot (BD - EB)/2 = 0$
$MA \cdot AD - MA \cdot EA + MB \cdot BD - MB \cdot EB = 0$
$MA \cdot AD + MB \cdot BD = MA \cdot EA + MB \cdot EB$: Conservation of Momentum

Derivation of Conservation of Kinetic Energy:

Taking the previous equations from Huygens's diagram and for the center of mass the following can be shown:

$MA \cdot EA^2 + MB \cdot EB^2 = MA \cdot (AD - 2AC)^2 + MB \cdot (BD - 2BC)^2$

$= MA \cdot AD^2 + MA \cdot 4AC^2 - MA \cdot AD \cdot 4AC + MB \cdot BD^2 + MB \cdot 4BC^2 - MB \cdot BD \cdot 4BC$

$= MA \cdot AD^2 + MB \cdot BD^2 + 4(MA \cdot AC^2 - MA \cdot AD \cdot AC + MB \cdot BC^2 - MB \cdot BD \cdot BC)$

$(MA \cdot EA^2 + MB \cdot EB^2)/(AC \cdot BC)$

$= (MA \cdot AD^2 + MB \cdot BD^2)/(AC \cdot BC) + 4((MA/BC) \cdot AC + (MB/AC) \cdot BC - (MA/BC) \cdot AD - (MB/AC) \cdot BD)$

$MA \cdot AC + MB \cdot BC = 0 \Rightarrow MA/BC = -MB/AC$

$\Rightarrow (MA \cdot EA^2 + MB \cdot EB^2)/(AC \cdot BC) = (MA \cdot AD^2 + MB \cdot BD^2)/(AC \cdot BC)$
$\qquad + 4((MA/BC) \cdot ((AC - BC) - (AD - BD)))$
$= (MA \cdot AD^2 + MB \cdot BD^2)/(AC \cdot BC)$
$\qquad + 4((MA/BC) \cdot ((AC + CB) - (AD + DB)))$
$= (MA \cdot AD^2 + MB \cdot BD^2)/(AC \cdot BC)$
$\qquad + 4((MA/BC) \cdot (AB - AB))$
$= (MA \cdot AD^2 + MB \: BD^2)/(AC \cdot BC)$

$\Rightarrow MA \cdot EA^2 + MB \cdot EB^2 = MA \cdot AD^2 + MB \cdot BD^2$: Conservation of *force vive*

Derivation of Conservation of Velocity:

$MA \cdot (AD^2 - EA^2) = MB \cdot (BD^2 - EB^2)$ *from conservation of force vive*
$MA \cdot (AD - EA) \cdot (AD + EA) = MB \cdot (BD - EB) \cdot (BD + EB)$
$MA \cdot (AD - EA) = MB \cdot (BD - EB)$ *from conservation of momentum*
$\Rightarrow AD + EA = BD + EB$: Conservation of velocity

4. Blackwell (1977: 581 note 8) interprets Huygens's quantity of motion as weight times velocity, which is not quite the same as Newton's mass times velocity. Stein (1990: 25) argues that Huygens had, indeed, anticipated Newton's distinction between mass and weight.
5. Stein (1990: 21–22) gives a very informative account of Huygens's impressive extension of relative motion arguments to generate his "wonderful law of nature," which anticipates Newton's corollary 4.

Rules of Impact and Reflection   189

6. A translation of this passage from Huygens's letter is also to be found in Hall 1966: 32.
7. In his *De motu corporum ex percussione*, published in the *Opuscula postuma* (1703), the propositions here numbered from 1 to 3 are labeled Hypothesis I, Hypothesis II, and Hypothesis III.
8. In the later published form of this treatise, this statement is preceded by the clause, "Whatever may be the cause of hard bodies rebounding from mutual contact when they collide with one another, let us suppose that . . ." As Blackwell notes (1977: 574 note 3), Huygens uses the word "hard" (*dura*) to indicate that his analysis is restricted to perfectly elastic collisions.
9. This hypothesis is absent from the treatise as published in 1703.
10. Huygens's demonstration of this proposition is given in *De motu corporum ex percussione*, Proposition 8; according to Blackwell (1977, note 1), this treatise was ready for publication in 1656, but was not published till 1703 in the *Opuscula postuma*.
11. See also Hall 1966: 32.
12. Blackwell (1977: 578, note 7) suggests that Huygens intends this proposition to be a direct refutation of Descartes's forth rule of impact. Here is the first sentence of the Millers' translation into English of sec. 49, Part II of Descartes's *Principles of Philosophy*:

> Fourth, if the body C were entirely at rest, <that is, if it not only had no apparent motion but also were not surrounded by air or any other fluid (which makes the hard bodies immersed in such a fluid very easily movable, as I shall show)>, and if C were slightly larger than B; the latter could never <have the force> to move C, no matter how great the speed at which B might approach.
>
> (Descartes 1991: 66)

13. As the symbolic rules make clear (see below), the amounts taken from R by the collision and added to S are each equal to the scalar sum of the amount by which the velocity of R was greater than its proper velocity together with the amount by which that of S was less than its proper velocity.
14. Hall (1966: 30–31) offers a translation of this first part of Wren's paper, but he omits Wren's symbolic rules for calculating these cases. These symbolic rules of Wren are included below in the translation by Wilson.
15. See Birch 1756–1757, 2: 335 and Hall 1966: 27.
16. See Hall 1966: 27.
17. See Clagett ([1959] 1961: Part I, Statics).
18. As Hall (1966: 29–30) points out, Wallis's model is the lever, where the product of forces and velocities is the same at either end.
19. See Mariotte 1673.
20. Proof that the velocity of the body is proportional to the chord of the arc TA for idealized pendulum motion without loss of energy:

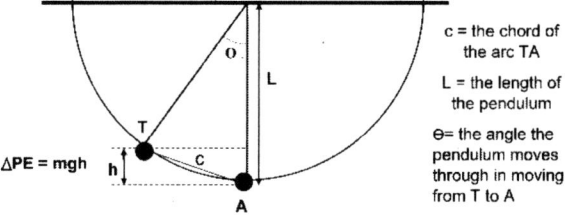

Figure 8.15 Pendulum arc velocity diagram.

This is modern proof using the conservation of total energy and the concepts of kinetic energy (KE) and potential energy (PE). At T the body is stationary, v = 0, it therefore has no kinetic energy. As the body is released from T it accelerates under the force of gravity to point A.

$\Delta PE + \Delta KE = 0$: *Conservation of energy*

In moving from point T to point A: $\Delta PE = -mgh$ and $\Delta KE = + mgh$

$KE = \frac{1}{2} mv^2 = mgh$ $\Rightarrow v = \sqrt{(2gh)}$.

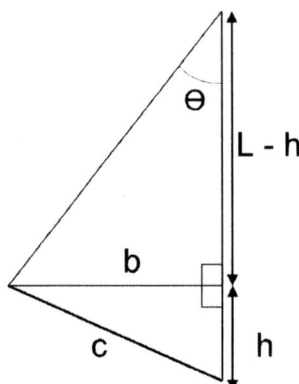

Figure 8.16   Euclid applied.

$c^2 = h^2 + b^2$
$L^2 = (L-h)^2 + b^2 = L^2 + h^2 - 2Lh + b^2$ $\Rightarrow b^2 = 2Lh - h^2$
$c^2 = h^2 + 2Lh - h^2$
$c^2 = 2Lh$ $\Rightarrow c = \sqrt{(2Lh)}$
$c = \sqrt{(2Lh)}$ $v = \sqrt{(2gh)}$
$c \: \alpha \: \sqrt{h}$ $v \: \alpha \: \sqrt{h}$ therefore $c \propto v$

21. Unlike imperfect elasticities, losses due to damage or deformation of the bodies would be difficult to measure directly. This, however, gives no reason to undermine our confidence in the equality of the action and reaction between bodies in such a collision.
22. It would have been easier to assess this if Newton had provided more details about data from actual trials of the experiments he described.

BIBLIOGRAPHY

Birch, T. (1756–1757) *The History of the Royal Society of London*, 4 vols, London.

Blackwell, R. J., trans. (1977) "Christiaan Huygens's *The Motion of Colliding Bodies*," *Isis*, 68: 574–597.

Bricker, P. and Hughes, R. I. G., eds. (1990) *Philosophical Perspectives on Newtonian Science*, Cambridge, MA: MIT Press.

Clagett, M. ([1959] 1961) The Science of Mechanics in the Middle Ages, Madison: University of Wisconsin Press.
Descartes, R. (1991) *Principles of Philosophy*, trans. V. R. Miller and R. P. Miller, Dordrecht: Kluwer.
Hall, A. R. (1966) "Mechanics and the Royal Society," *British Journal for the History of Science*, 3: 24–38.
Huygens, C. (1669) "A Summary Account of the Laws of Motion," *Philosophical Transactions*, 4: 925–928.
———. (1703) *Opuscula postuma*, Leiden.
———. (1888–1950) *Oeuvres complètes de Christiaan Huygens*, Pubiées par la société hollandaise des sciences, 22 vols, The Hague: Martinus Nijhoff.
Mariotte, E. (1673) *Traité de la percussion ou chocq des corps, dans lequel les principales règles du mouvement contraines a celles que M. Descartes, & quelques autres modernes ont voulu establir, sont démontrées par leurs véritables causes*, Paris.
Newton, Sir I. (1999) *The Principia: Mathematical Principles of Natural Philosophy*, eds. and trans. I. B. Cohen and A. Whitman, Berkeley: University of California Press; 1st edn 1687.
Stein, H. (1990) "On Locke, 'the Great Huygenious, and the incomparable Mr. Newton'" in eds. P. Bricker and R. I. G. Hughes 1990, pp. 17–47.
Wallis, J. (1668) "A Summary account by Dr. John Wallis, of the General Laws of Motion, by way of letter written by him to the Publisher, and communicated to the R. Society," *Philosophical Transactions*, 3: 864–866.
Wren, C. (1668) "Dr. Christopher Wren's Theory concerning the same subject 'The Law of Nature concerning Collision of Bodies,'" *Philosophical Transactions*, 3: 867–868.

Part IV
Leibniz and Hume

9 Leibniz, Body, and Monads

Daniel Garber

Everyone knows that for Leibniz, the ultimate make-up of the world is monads. As he puts it at the beginning of the *Monadology*:

> The monad, which we shall discuss here, is nothing but a simple substance that enters into composites—simple, that is, without parts. . . . But where there are no parts, neither extension, nor shape, nor divisibility is possible. These monads are the true atoms of nature and, in brief, the elements of things.
>
> (*Monadology* §§ 1, 3; AG 213)

Monads are mind-like, nonextended, like Cartesian souls of a sort. This seems like a very idealistic view of the world. But what about bodies? Where exactly do bodies fit into Leibniz's world?

This is my project in this chapter, to find the place to which bodies vanished in Leibniz's monadological metaphysics.

BEFORE MONADS

Before addressing my main question, I would like to begin by examining Leibniz's philosophy before he came to the theory of monads for which he is best known. Contrary to much of the literature, Leibniz did not always have the problem of bodies in the context of monads—because he did not always have monads.

This is not the place for a full account of the development of Leibniz's views on body.[1] Leibniz was concerned with body and its metaphysical foundation from his earliest years. The early years are complicated, but by the 1680s, Leibniz had attained a kind of stability in his views, and articulated a beautiful picture of the way the world is, a subtle and complex conception of body and its make-up.

The view is articulated most clearly in the correspondence Leibniz had with Antoine Arnauld in 1686 and 1687, a correspondence that he had contemplated publishing in later years. In his early years, Leibniz had been attracted to something like the Cartesian view of body, though Hobbes was probably more of a direct influence than Descartes himself. Descartes had argued that bodies were extended and extended alone, the objects of geometry made real. In that way, his intention was to eliminate the substantial forms that were central to the conception of body and substance among the scholastic Aristotelians.[2] But a crucial innovation came when Leibniz decided that extension was not enough: we have to rehabilitate substantial forms. As he wrote to Arnauld, himself a Cartesian on this issue:

> If the body is a substance and not a simple phenomenon like the rainbow, nor an entity united by accident or by aggregation like a heap of stones, it cannot consist of extension, and one must necessarily conceive of something there that one calls substantial form, and which corresponds in a way to the soul. I have been convinced of it finally, as though against my will, after having been rather far removed from it in the past.
> (Leibniz to Arnauld, July 4/14, 1686, A2.2.82 (G II 58))[3]

An important consideration for Leibniz here is unity: for something to exist, it must have genuine unity. As he writes to Arnauld:

> I believe that where there are only entities through aggregation, there will not even be real entities; for every entity through aggregation presupposes entities endowed with a true unity.... I do not grant that there are only aggregates of substances. If there are aggregates of substances, there must also be genuine substances from which all the aggregates result. One must necessarily arrive either at mathematical points from which certain authors make up extension, or at Epicurus's and M. Cordemoy's atoms (which you, like me, dismiss), or else one must acknowledge that no reality can be found in bodies, or finally one must recognize certain substances in them that possess a true unity.
> (Leibniz to Arnauld, April 30, 1687, A2.2.184–185 (G II 96))

And so, he argues, the basic elements of the physical world are corporeal substances: unities of matter and form. Again, as he wrote to Arnauld:

> I accord substantial forms to all corporeal substances that are more than mechanically united.... If I am asked for my views in particular on the sun,... the earth, the moon, trees and similar bodies, and even on animals, I cannot declare with absolute certainty if they are animate or at least if they are substances or even if they are simply machines or aggregates of many substances.... [E]very part of matter is actually divided into other parts as different as the diamonds [of the Grand Duke and the Grand Mogul]; and since it continues endlessly in this way, *one will never arrive at a thing of which it may be said: "Here really*

is an entity," except when one finds animate machines whose soul or substantial form creates substantial unity independent of the external union of contiguity. And if there are none, it follows that apart from man there is apparently nothing substantial in the visible world.
  (Leibniz to Arnauld, November 28/December 8, 1686, A2.2.121–122 (G II 77))

These corporeal substances, these unities are, in turn, made up of smaller corporeal substances, and so on to infinity. As Leibniz writes about human beings:

> ... man ... is an entity endowed with a genuine unity conferred on him by his soul, notwithstanding the fact that the mass of his body [*la masse de son corps*] is divided into organs, vessels, humours, spirits, *and that the parts are undoubtedly full of an infinite number of other corporeal substances endowed with their own entelechies.*
> (Leibniz to Arnauld, October 9, 1687, A2.2.251 (G II 120))[4]

Or, Leibniz writes about corporeal substances in more general terms:

> If one considers the matter of the corporeal substance not mass without forms but a secondary matter which is the multiplicity of substances of which the mass [*masse*] is that of the entire body, it may be said that these substances are parts of this matter, just as those [substances] which enter into our body form part of it, for as our body is the matter and the soul is the form of our substance, it is the same with other corporeal substances.
> (Leibniz to Arnauld, October 9, 1687, A2.2.250 (G II 119))[5]

This, then, is what grounds reality, what are the ultimate elements of things for Leibniz in this period, corporeal substances, unities of form and matter, soul and body that, in turn, contain corporeal substances smaller still, to infinity. Because of this, at every level we have both unities while, at the same time, we have unities at a lower level, infinite divisibility without having to give up unity.

It is uncontroversial that Leibniz did not use the term "monad" until the mid-1690s. It first appears in a letter to L'Hospital dated July 12/22, 1695, and begins to appear more frequently in the late 1690s.[6] But there is some debate about whether Leibniz's metaphysics during the 1680s was essentially monadological. That is, many commentators hold that even in the period of the correspondence with Arnauld, the ultimate foundation of the world for Leibniz was nonextended and mind-like entities, what he will later call monads.[7] I do not think that there is any good reason to believe that the monadological metaphysics that is so prominent in his later writings can be found before the mid-1690s at least.[8] But by the period of the *New*

System (1695), Leibniz begins to contemplate adding monads to his metaphysics as a subbasement to ground the physical world; by 1700 or so, the new metaphysics is clearly there. When I say that Leibniz "added" monads as a subbasement, my words are carefully chosen. The world of bodies, in a way, is still there, and Leibniz does not mean to eliminate it, I would argue. But Leibniz no longer believes that bodies can stand alone, as it were: they require a foundation, and it is monads that are supposed to give bodies a foundation. Nor is it entirely clear exactly why Leibniz wants to add this new metaphysical level; unfortunately he never addresses the question directly. One obvious reason has to do with an argument quoted above. In the 1680s Leibniz wrote over and over again that for anything to be real, it must be grounded ultimately in genuine unities: "I believe that where there are only entities through aggregation, there will not even be real entities; for every entity through aggregation presupposes entities endowed with a true unity. . . ."[9] Now, in the correspondence with Arnauld and other writings from this period, Leibniz seems perfectly happy to grant that corporeal substances, human beings, for example, can be regarded as genuine unities in this sense. But by the mid- and late 1690s, he seems no longer to accept such a position. Here is what he wrote to a correspondent, Johann Gebhard Rabener, in January (?) 1698:

> . . . since an aggregate is constituted by simples, I later discovered that we must arrive at monads. Not, indeed, corporeal or spatial [monads], since the continuum is not composed of indivisibles, nor are there any material atoms, but, however, substantial [monads]. Therefore every true monad is a simple substance, and is in some sense analogous to a mind, and that hence it follows that [every monad] is coeval with the world, unless it was created by God in the course of time.
>
> (A1.15.260)[10]

I hesitate to say that Leibniz had definitely adopted the monadological metaphysics by this point; there are other texts from the same period and after where he seems to take quite a different view.[11] But it is quite clear that by the late 1690s, the view is on the table.

However, with this new metaphysics begins the problem of bodies. With the world of monads introduced now as "the true atoms of nature and . . . the elements of things," we seem to have a new metaphysics and a new question to answer: how are these metaphysical atoms related to bodies? What becomes of bodies in the new metaphysics?

MONADS AND BODIES

There are a number of strands in Leibniz's thought about the relation between monads and bodies in his later monadological metaphysics. This is not a discussion that takes place in a public forum, where monads themselves are hardly in view. However, the question is very much in evidence in two of Leibniz's most important sets of correspondence, the correspondence with de Volder, and the later correspondence with Des Bosses. Burcher de Volder (1643–1709) was a professor of physics at Leiden, a Cartesian at heart, though he was impressed by English experimental philosophy. Bartholomaeus Des Bosses (1668–1728) was a Jesuit mathematician, later to become the Latin translator of Leibniz's *Theodicy*. In his correspondences with these two very different interlocutors, Leibniz discussed the notion of body and its relation to monads at great length.

Strand 1: Bodies as Aggregates of Monads

On one strand of Leibniz's thought, bodies are to be understood as aggregates of monads. This is strongly suggested in the opening of the *Monadology*, quoted above.

It is even clearer in the opening of the *Principles of Nature and Grace*:

> A substance is a being capable of action. It is simple or composite. A *simple substance* is that which has no parts. A *composite substance* is a collection of simple substances, or *monads*. *Monas* is a Greek word signifying unity, or what is one. Composites or bodies are multitudes; and simple substances—lives, souls, and minds—are unities. There must be simple substances everywhere, because, without simples, there would be no composites.
> (*Principles of Nature and Grace* § 1, AG 207)[12]

But the view is also quite clearly found in earlier texts. In his letter to de Volder from Winter 1703, sent again on June 20, 1703, he wrote that " ... simple things alone are true things, the rest are only beings through aggregation, and therefore phenomena, and, as Democritus used to say, exist by convention not in reality" (G II 252),[13] a view that he repeated a number of times in that correspondence.[14]

Leibniz's correspondence with de Volder ended in January 1706. But almost immediately after Leibniz ended that exchange, he began exchanging ideas with Des Bosses. The question of body comes up early in the exchange, and becomes one of the central questions that the two discuss. And in one of Leibniz's first letters to Des Bosses, he presents the view of bodies as aggregates of monads:

> ... from many monads there results secondary matter, together with derivative forces, actions, and passions, which are only beings through aggregation, and thus semi-mental things, like the rainbow and other well-founded phenomena.
> (Leibniz to Des Bosses, March 11, 1706, LDB 34)[15]

Later in the correspondence, Leibniz will introduce his difficult and much-discussed notion of the substantial bond, the *vinculum substantiale* that is supposed to transform aggregates of monads into genuine corporeal substances. We will shortly discuss this notion and the account of corporeal substances Leibniz offers in these texts. With the substantial bond there emerges a very different conception of what bodies are and how they are related to monads. But, as we shall see, it is not always clear whether Leibniz meant to adopt the hypothesis. In numerous places, he contrasts what the world of bodies would be like with and without the hypothesis of substantial bonds. And in at least some such texts, Leibniz suggests that without substantial bonds, bodies would be aggregates of monads:

> If you deny that what is superadded to monads in order to make a union is substantial, then body cannot be said to be a substance, for in that case it will be a mere aggregate of monads, and I fear that you will fall back on the mere phenomena of bodies.
> (Leibniz to Des Bosses, May 26, 1712, LDB 240)[16]

And two years after this and related texts, we find the aggregate view of bodies expressed in the *Monadology*, the *Principles of Nature and Grace* and related texts.

Problems with the Aggregate View

But there are some obvious problems with the view of bodies as aggregates of monads. The most obvious problem regards their extension: how can an aggregate of nonextended substances be extended? This is something that disturbed Des Bosses:

> I do not yet understand from the things you have said so far—either because I have not sufficiently penetrated them or because you assume some principle that is unknown to me—how mass, which is real and has a real diffusion or extension, can result from monads alone, which lack diffusion and extension.
> (Des Bosses to Leibniz, September 6, 1709, LDB 146)[17]

Even if we can understand how extension might arise from the perception of an infinity of nonextended monads, it is still a bit puzzling as to how we can understand the claim that we are perceiving monads at all! Furthermore, Leibniz asserts again and again that not only is extension phenomenal, but so are forces, active and passive. In reinterpreting the world in terms of monads, the derivative active and passive forces that had earlier been real parts of a real physical world—modes of corporeal substances—are now taken to be phenomenal: "I relegate derivative forces to the phenomena" (Leibniz to de Volder, January 1705, G II 275). What exactly does Leibniz mean by calling derivative forces phenomenal?

Strand 2: Common Dream View

In addition to the view of bodies as aggregates, one can find at least one other conception of the relation between monads and bodies that features prominently in Leibniz's writings in this period. This view is suggested in some passages from a letter that Leibniz wrote to Des Bosses on June 30, 1704:

> Indeed, considering the matter carefully it should be said that there is nothing in things except simple substances and in them perception and appetite. Moreover, matter and motion are not so much substances or things as the phenomena of perceivers, the reality of which is located in the harmony of perceivers with themselves (at difference times) and with other perceivers.
>
> (G II 270)

Understood in this way, it would appear that bodies are to be understood as the contents of the perceptions of individual monads, which are coherent within individual monads, and coherent from one monad to another. Elsewhere in the same letter he writes:

> But accurately speaking, matter is not composed of constitutive unities, rather it results from them, since matter or extended mass is nothing but a phenomenon founded in things, like the rainbow or the perihelion. And there is no reality in anything except the reality of unities, and so phenomena can always be divided into lesser phenomena that could appear to other more subtle animals, and the smallest phenomena will never be reached. By contrast, substantial unities are not parts, but the foundations of the phenomena.
>
> (G II 268)

This passage also suggests the kind of phenomenalism to which Leibniz alludes in the first passage quoted, where the bodies are conceived of as the common dream of an infinity of monads. In this view, monads are no longer constituents of bodies in any real sense: they are what ground the existence of bodies, presumably through their perceptions of them. What is particularly striking here is the distance between the world of phenomena and the world of substances. Phenomena "can always be divided into lesser phenomena": there is no bottom level of phenomena. But existing on another plane, as it were, are the monads, whose perceptions ground the world of phenomena.

This strand is largely in the background of the exchange with de Volder. But in some of the later letters to Des Bosses, it becomes considerably more prominent, and emerges as an account of what the world would be like if there were no substantial bonds and thus no true corporeal substances. In the letter Leibniz wrote to Des Bosses on February 15, 1712, the letter that first introduces the notion of a substantial bond, Leibniz writes:

> If that substantial bond of monads were absent, then all bodies with all their qualities would be only well-founded phenomena, like a rainbow or an image in a mirror, in a word, continuous dreams that agree perfectly with each other; and in this alone would consist the reality of those phenomena. For it should no more be said that monads are parts of bodies, that they touch each other, that they compose bodies, than it is right to say this of points and souls.
> (LDB 226)

The view is presented in a way that is even starker in a letter written a few months later:

> It is true that the things that happen in the soul must agree with those which happen outside the soul; but for this it is sufficient that those things that happen in one soul correspond both among themselves and with those things that happen in any other soul; and there is no need to posit something outside of all souls or monads. According to this hypothesis, when we say that Socrates is sitting, nothing more is signified than that those things that we understand by "Socrates" and "sitting" are appearing to us and to others for whom it is a concern.
> (Leibniz to Des Bosses, June 16, 1712, LDB 256)

Something very like this view can be found in the very last letter that Leibniz wrote to Des Bosses, just months before he died:

> Moreover, if monads alone were substances, one of two things would be necessary: either bodies would be mere phenomena or a continuum would arise from points, which we agree is absurd. Real continuity can arise only from a substantial bond. If there existed nothing substantial besides monads, that is, if composites were mere phenomena, then extension itself would be nothing but a phenomenon resulting from coordinated simultaneous appearances, and by that fact, all the controversies concerning the composition of the continuum would cease.
> (Leibniz to Des Bosses, May 16, 1716, LDB 370)

It is, admittedly, not altogether clear what is going on in this passage. But the "coordinated simultaneous appearances" to which Leibniz refers here might well be the "continuous dreams that agree perfectly with each other" that are at issue in the earlier letter.[18]

Strengths and Weaknesses of the Common Dream View

This view has some definite attractions over and above the view of bodies as aggregates of monads. For example, it is no longer a problem from where extension is supposed to derive. Extension is purely phenomenal, and appears *only* in the contents of monadic perceptions. It is a feature of the common dreams monads have, but there is no pretense that it is anything more than that. Similarly, there is now a simple and coherent account of force, active and passive. All forces, at least bodily forces, now pertain directly to phenomena: they are the phenomenal properties of phenomenal bodies, and nothing more. Exactly this question is at issue in a remark that Leibniz added to a letter to Bourguet, March 22, 1714:

> The difficulty that one has about the communication of motion ceases when one considers that material things and their motions are only phenomena. Their reality is only in the agreement of the appearances of monads. And if the dreams of one single person were entirely coherent and if the dreams of all souls agreed with one another, one wouldn't concern oneself with anything else in order to make of them body and matter.
>
> (G III 567)

The resolution of this problem comes at something of a cost, though: the world would have the same metaphysical status as a dream.

MONADS, BODIES, AND CORPOREAL SUBSTANCE

I have been trying to pin Leibniz down on the question of the nature of extended bodies within the context of the monadology. But there is another question, not altogether unrelated to this one. For Leibniz in this period, monads are true substances, the elements of things, in whatever sense he understands this locution. But are these the *only* substances that Leibniz recognizes? Are there no corporeal substances left in Leibniz's world? In particular, how do *bodies* fit into the world of *substances*?

Bodies Not Substances

Here, again, there seem to be a number of strands in Leibniz's thought. There are a number of striking passages in which Leibniz seems to hold that monads or simple substances alone are true substances, implying that he recognizes no other substances. Writing to Sophie on November 19, 1701, Leibniz discussed true unities, which have neither parts nor extension. In this context he claimed that "there is nothing real but the unities" and that "everything is unities" (A1.20.74–75). These views are repeated a number of times in the letters to de Volder. In the letter from Winter 1703, sent again on June 20, 1703 he writes that:

> ... simple things alone are true things, the rest are only beings through aggregation, and therefore phenomena, and, as Democritus used to say, exist by convention not in reality.
> (G II 252)

A year later he wrote in an often-quoted passage that "... considering the matter carefully it should be said that there is nothing in things except simple substances and in them perception and appetite" (Leibniz to de Volder, June 30, 1704, G II 270). Writing in January 1705, he is blunter still. De Volder, struggling to make sense of Leibniz's view, complained:

> It now seems to me that you do away with bodies altogether, in as much as you place them only in appearances, and that you substitute forces alone for things; and not even corporeal forces, but "perception and appetite."
> (De Volder to Leibniz, November 14, 1704, G II 272)

Leibniz replies:

> I do not really do away with body, but reduce it to what it is. For I show that corporeal mass that is believed to have something besides simple substances, is not a substance, but a phenomenon resulting from simple substances, which alone have unity and absolute reality. I relegate derivative forces to the phenomena, but I think that it is clear that primitive forces can be nothing other than the internal strivings of simple substances, by which they pass from perception to perception by a certain law of their nature and at the same time agree with each other. ... It is necessary that these simple substances exist everywhere and that they are self-governing (each as far as itself is concerned), since the influence of one on another cannot be understood. Anything more beyond this in things is posited in vain and added without argument. ... Whoever adds anything to these things will accomplish nothing, and will both work in vain in giving explanations and be thrown into inextricable difficulties. ... Indeed, I suppose nothing everywhere and throughout all things except that which we all admit in our own souls on many occasions, namely, internal spontaneous changes. And in this way I exhaust the totality of things with one act of the mind.
> (Leibniz to de Volder, January 1705, G II 275–276)[19]

In this way, Leibniz seems to hold, all there really is in the world are the monads.

While this view is clearest and presented most forcefully in the writings from the early part of the first decade of the eighteenth century, the view persists in a number of texts, for example in a letter to another correspondent, Pierre Dagnicourt, on September 11, 1716:

> I am also of the opinion that, to speak exactly, there is no need of extended substance. . . . True substances are only simple substances or what I call "monads." And I believe that there are only monads in nature, the rest being only phenomena that result from them. Each monad is a mirror of the universe according to its point of view and is accompanied by a multitude of other monads which compose its organic body, of which it is the dominant monad.
> (Leibniz 1734: 1–2 (LDB 401n105))

Here, a month before he died, Leibniz seems to deny that we need anything over and above monads.

Bodies as Composite Substances

But alongside these texts, there are others in which Leibniz seems to advance the view that beside monads or simple substances there are also compound or composite substances. There are a number of texts that are very suggestive of the view of corporeal substances found, for example, in the correspondence with Arnauld. For example, Leibniz wrote the following in a letter to Bierling, August 12, 1711:

> Finally, you ask for definitions of matter, body, and spirit. Matter is that which consists in impenetrability [*Antitypia*], or that which resists penetration, and therefore bare matter is merely passive. Moreover, over and above matter, body also has active force. Furthermore, a body is either a corporeal substance or a mass composed of corporeal substances. I call a corporeal substance that which consists of a simple substance or monad (that is, a soul or something analogous to a soul) with an organic body united to it. But mass [*Massa*] is an aggregate of corporeal substances, as a cheese consists in a coming together of worms. Furthermore, a monad or simple substance in general contains perception and appetite.
> (G VII 501–502)

In this and a number of similar texts from the period,[20] it looks as if Leibniz is just reprising his earlier view of corporeal substances, souls that give unity to their organic bodies and transform them from mere aggregates whose unity is imposed on them by the perceiver into genuine substances.

But there are other texts that suggest somewhat different conceptions of composite substance. This view is also suggested in the opening paragraph of the *Principles of Nature and Grace*, in a passage we saw earlier in connection with the monadic aggregate view of body. There Leibniz writes:

> A substance is a being capable of action. It is simple or composite. A *simple substance* is that which has no parts. A *composite substance* is a collection of simple substances, or *monads*. *Monas* is a Greek word signifying unity, or what is one. Composites or bodies are multitudes; and simple substances—lives, souls, and minds—are unities. There must be simple substances everywhere, because, without simples, there would be no composites. As a result, all of nature is full of life.
> (*Principles of Nature and Grace* § 1, AG 207)

Here the view seems to be that the composite substance is just a collection of monads, united by a single monad "which makes up the center" and that is "the principle of unity" for that composite.[21]

This conception of composite substance is importantly different from the conception that he held earlier, in the correspondence with Arnauld, for example. In the earlier conception, there seemed to be a robust sense of body and matter at issue in the composite: the bodies that the form or soul was supposed to make into one were supposed to be the seat of passive force, where passive force was understood as involving resistance and impenetrability. While these bodies are not formally extended in the Cartesian sense, their passive force is what gives rise to extension. However, in this later view, the bodies transformed into composite substances are themselves just aggregates of nonextended and mind-like simple substances. While each simple substance has passive force in addition to matter, the passive force here is understood not as resistance and impenetrability but as a feature of the internal states of the monads in question, their confused perception or imperfection.

Substantial Bonds

But there is another approach to composite substance in Leibniz's texts, one that goes beyond the older model of corporeal substance in the correspondence with Arnauld, and the monads-together model that appears to replace it in some later texts: the theory of the substantial bond that Leibniz developed in the correspondence with Des Bosses.[22]

A reasonably full version of the substantial bond theory is outlined in the letter where it is introduced, and many of the details are filled out in letters written over the following year. Leibniz begins the relevant passage of his letter of February 15, 1712, as follows:

> If corporeal substance is something real over and above monads, as a line is taken to be something over and above points, we shall have to

say that corporeal substance consists in a certain union, or rather in a real unifier superadded to monads by God, and that from the union of the passive powers of monads there in fact arises primary matter, which is to say, that which is required for extension and antitypy, or for diffusion and resistance. From the union of monadic entelechies, on the other hand, there arises substantial form; but that which can be generated in this way, can be destroyed and will be destroyed with the cessation of the union, unless it is miraculously preserved by God. But such a form then will not be a soul, which is a simple and indivisible substance. And this form, just like matter, is in perpetual flux, since in fact no point can be designated in matter that preserves the same place for more than a moment and does not move away from neighboring points, however close. But a soul in its changes persists as the same thing, with the same subject remaining, which is not the case in a corporeal substance. Thus, one of two things must be said: either bodies are mere phenomena, and so extension also will be only a phenomenon, and monads alone will be real, but with a union supplied by the operation of the perceiving soul on the phenomenon; or, if faith drives us to corporeal substances, this substance consists in that unifying reality, which adds *something absolute* (and therefore substantial), albeit impermanent, to the things to be unified.

(LDB 224–226)

What is at issue here seems to be the union of monads into a single substance; the substantial bond is "a real unifier superadded to monads by God." But not just any collection of monads can be united. In a related text that Look and Rutherford present as a supplement to the letter of February 15, 1712, Leibniz writes:

This addition to monads does not occur in just any way, otherwise any scattered things at all would be united in a new substance, and nothing determinate would arise in contiguous bodies. But it suffices that it unites those monads that are under the domination of one monad, that is, that make one organic body or one machine of nature.

(LDB 232)

What is added is something that is not itself just another monad, but yet it is a something, something substantial rather than a mode of the monads, "*something absolute* (and therefore substantial), albeit impermanent, [added] to the things to be unified."

Leibniz is somewhat vague about the relations between the monads and the substantial bond that unites them. In some sense the substantial bond seems to reflect that perceptions and appetitions of the monads to which it is connected, while, in a certain sense, being independent of them. In his letter to Des Bosses from January 24, 1713, Leibniz wrote:

> A substantial bond superadded to monads is, in my opinion, something absolute, which although it accurately corresponds in the course of nature to the affections of monads, namely their perceptions and appetitions, so that in a monad it can be read in which body its body is, nevertheless, supernaturally, the substantial bond can be independent of the monads and can be changed and accommodated to other monads, with the previous monads remaining.
>
> (LDB 296)

As the correspondence goes on, the substantial bond seems to get closer and closer to being the corporeal substance itself. This view is first articulated in notes connected with the letter of January 24, 1713, that the substantial bond *just is* the corporeal substance: "... I do not know what a being realizing phenomena is except that very thing that I call a composite substance or substantial bond" (LDB 304). In the letter of August 23, 1713 Leibniz continues this thought, referring to "corporeal substance, or the substantial bond of monads." By the end of the series of letters, the corporeal substance and the substantial bond seem to be completely identified with one another. In the letter of January 13, 1716, Leibniz writes:

> I do not see how it can be conceived that the thing realizing the phenomena is something apart from substance. For that realizing thing must bring it about that composite substance contains something substantial besides monads, otherwise there will be no composite substance, that is, composites will be mere phenomena. And in this I think that I am absolutely of the same opinion as the Scholastics, and, in fact, I think that their primary matter and substantial form, namely the primitive active and passive powers of the composite, and the complete thing resulting from these, are really that substantial bond that I am urging.
>
> (LDB 364)

And then, in the final letter he wrote to Des Bosses on May 29, 1716, Leibniz says something absolutely remarkable:

> Composite substance does not consist formally in monads and their subordination, for then it would be a mere aggregate, that is, an accidental being; rather, it consists in primitive active and passive force, from which arise the qualities, actions and passions of the composite, which are perceived by the senses, if more than phenomena are assumed.
>
> (LDB 371)

The substantial bond now looks like the corporeal substance of the Arnauld correspondence, resurrected, and beginning to shove the monads aside: it is not at all clear what work they are now doing in the theory. The substantial bond, originally introduced to bind the monads together into a composite substance, has largely taken over. In this last view, it is not clear exactly what work monads are supposed to do; they seem almost dispensable.

I think that it is very unlikely that Leibniz would have rested with the view that finally emerges at the end of the correspondence with Des Bosses: the question of the coherence of the view aside, it is implausible that he could at this stage have given up on monads. But at the same time, it seems clear to me that in these letters, Leibniz is genuinely worried about the reality of body and the physical world, and is experimenting around with a way of reviving his earlier and more robustly realistic view that will mesh with the new monadology. He seems willing to entertain giving up the radical monads only metaphysics that he initially espoused in the letters with de Volder—and to introduce something new in nature over and above the monads—in order to save the reality of bodies. But I suspect that he had a hunch that he had not quite solved the problem yet.

WHAT DOES IT ALL MEAN?

In this chapter we have been examining Leibniz's treatment of body in the context of a world grounded in monads. What, in the end, was Leibniz's view? Leibniz's account of body in these years does not make for a fully coherent picture. There are texts that suggest that bodies are aggregates of monads, and others that suggest that bodies are just the common coherent dreams of an infinity of monads. There are texts that suggest that all there are are monads in the world, and that everything else is just phenomena, while other texts suggest that there are, in addition, composite or corporeal substances. One can even find a variety of different conceptions of what these composite substances are supposed to be in these years. In some texts, they seem to be the familiar corporeal substance of his earlier years, souls or forms that transform organic machines into genuine unities, however those organic machines may be understood; in others the composite substances seem to be aggregates of monads united by virtue of a dominant monad; in yet others, the unity requires something that goes beyond the ontology of monads, a substantial bond, a strange entity that comes out looking more and more like an old-fashioned corporeal substance the more Leibniz looked at it. And while there are some temporal patterns that can be seen by the discerning eye—the career of the substantial bond, for example, which only seems to enter fairly late in the game—many of these positions seem to be mixed in no discernable order from 1700 or so, when Leibniz first seems to sign on to the monadological metaphysics, to the time of his death in 1716. They seem to come and go, bob and weave through Leibniz's thought, as if Leibniz did not even recognize that they are incompatible. What are we to make of this?

Here is an attempt to make at least some limited sense of the apparent chaos. One constant in the whole period, from 1700 or so to the end of Leibniz's life, is the theory of monads: in one way or another, monads are to be understood as the ultimate elements of things, the bottom layer in our account of the way the world is. This is, of course, a metaphysical claim. But we should not misunderstand what that would have meant for Leibniz. It may sound strange to put it this way, but in some ways what Leibniz was doing is more like contemporary particle physics than it is like contemporary analytic metaphysics. It is not unfair to say that Leibniz uses metaphysical arguments to limn the ultimate nature of the physical world in the way in which a contemporary physicist might use mathematics or symmetry arguments for the same purpose. But putting it in that way distorts the project as well. What Leibniz is doing is a kind of enterprise that we do not do today, either in physics or in philosophy: it is (natural) philosophy as Leibniz and his contemporaries understood the enterprise.[23]

How does this help make sense of Leibniz's thought? A modern physicist may have good reason to believe that ultimate reality is quite different than what we experience around us: it may obey very different laws than macroscopic objects do, and violate constraints and regularities that we take for granted about the physical world in which we live. If something like string theory turns out to be true, the ultimate reality may be *very* different from anything that we experience. But yet, the physicist must try to figure out just how this fundamental physics can give rise to what we see around us. This is the problem, familiar from the philosophy of science, of the relation between the scientific and manifest image, the world as it appears to the scientist, and the world as it appears to our everyday experience. Leibniz has a similar problem. One of Leibniz's commitments in his later years was to the world of nonextended and mind-like monads as the foundations of the real world. But Leibniz did not want to let go of the reality of the world of bodies that we experience either. And this leads to a question: how is it that the world of our everyday experience is related to what he takes to be the fundamental metaphysics, the world of monads? Certain metaphysical arguments convinced Leibniz that at root, simple substances, monads had to be at the bottom of everything. What he had not fully figured out, though, is how exactly the world of bodies is grounded in the world of monads. This, I would claim, was the project of the letters with de Volder, the letters with Des Bosses, and other texts of this period. And, I would claim, there is no single doctrine that you can say is *the* Leibnizian solution to that problem. There are different strands that recur throughout the texts, but I do not think that he ever arrived at an answer that fully satisfied him.

It is because of this that it is difficult to offer a neat and clean characterization of what exactly the monadological metaphysics comes to, or to offer a neat and clean comparison with the earlier corporeal substance view. The earlier corporeal substance view is clear enough, but the monadological metaphysics is a kind of moving target, very difficult to pin down, since an essential piece of it, the relation between the metaphysically foundational world of monads and the experientially fundamental world of bodies, is constantly up for grabs. The best that you can do is say that in the new view, monads are the ultimate elements of things, as opposed to the corporeal substances of the earlier view. But not even this will do, since by the end, by the substantial bond view in the last letters with Des Bosses, even the monads seem to be called into question.

It also makes it difficult to determine whether or not Leibniz was an idealist in his later years. For generations, historians of philosophy have assumed that Leibniz was an idealist. One can point to texts that, it would seem, could *only* be interpreted as idealistic in a robust sense: "Indeed, considering the matter carefully it should be said that there is nothing in things except simple substances and in them perception and appetite" (Leibniz to de Volder, June 30, 1704, G II 270). Recent commentators have challenged this in various ways.

But I wonder about the whole discussion. "Realist" and "idealist" are just not terms that Leibniz himself used to describe his position. Nor were they terms that were really available to him; as terms of art in philosophical discourse, they do not really enter the vocabulary until after Leibniz's death.[24] I wonder if this is not just a bad question to be focusing on.[25] For this reason I have avoided the term in earlier discussions in this chapter. Leibniz, in the end, wants to maintain the importance of monads as a foundation for everything, at least in the monadological years after 1700 or so, and up until practically the end of the exchange with Des Bosses. And he never gives up the idea of a world of bodies ultimately grounded in living things, bugs in bugs to infinity. And he wants to figure out how to relate the two to one another. Here he is simply unsure how to go. Sometimes he gives more weight and heft to the bodies; when he does, we call him a realist; sometimes he gives more reality to the monads, and when he does we call him an idealist. But at least in his later years, both monads and bodies are always there in one way or another. It is, perhaps, open to us to represent him as swinging from one pole to another on the metaphysical scheme of things. Or we can just see it for what it is, Leibniz in his metaphysical workshop, trying out different ways of connecting the two pieces of his world, both of which must find their places in his final story.

Historians of philosophy want to find Leibniz's "considered view"; much of the debate in this area comes down to differences about where different commentators come down on this question. But I am not sure that there is a considered view to find. In his later years, in my reading, Leibniz was struggling with the problems of bodies, substances, and monads, and struggling toward a considered view on these issues. But he died before he got there.

NOTES

1. For a fuller account of the development of Leibniz's ideas about body, see Garber 2009. This chapter summarizes some of the main themes from that book.
2. For more on this, see Garber 1992, Ch. 4.
3. References to the Leibniz/Arnauld correspondence are given both in the current Akademie edition volume, recently published, and in the more widely

cited but older version in Gerhardt. Translations are from Leibniz and Arnauld 1967, which is keyed to the pagination in Gerhardt.
4. Cf. Leibniz to Malebranche, June 22/July 2, 1679, A2.1^2.719.
5. The editors signal this as a later edition to the text.
6. See A3.6.451 (WF 57). See Garber 2009, Ch. 8, for a fuller discussion of the introduction of the term into Leibniz's vocabulary in the late 1690s.
7. See especially Adams 1994, part III, and Sleigh 1990, Ch. 5 and 6, who argue explicitly for this view.
8. The full case for this claim is made in Garber 2009.
9. Leibniz to Arnauld, April 30, 1687, A2.2.184 (G II 96).
10. In the Vorausedition for A2.3, the Akademie editors recently published a fascinating exchange with Gabriel Wagner from December 1697 and March 1698 where Leibniz seems clearly to accept nonextended monads as foundational. The texts can be downloaded from their website: http://www.leibniz-edition.de/Baende/ReiheII.htm.
11. For example, in a letter to Johann Bernoulli from September 20/30, 1698, later in the same year as the text he wrote to Rabener and Wagner, Leibniz wrote: "What I call a complete monad or individual substance [*substantia singularis*] is not so much the soul, as it is the animal itself, or something analogous to it, endowed with a soul or form and an organic body" (GM III 542 (AG 168)).
12. See also the text that Leibniz had originally written for Remond in 1714 to explain his views: "I believe that the entire universe of creatures consists only in simple substances or monads, and in their aggregates. These simple substances are what we call mind in us and in spirits, and soul in animals.... Aggregates are what we call bodies" (G III 622).
13. I am deeply indebted to Paul Lodge, who shared with me his new edition and translation of the letters between Leibniz and de Volder, as well as his deep knowledge of the exchange. All of the translations I use of the de Volder texts are his. Lodge's translations will soon appear in the *Yale Leibniz*.
14. See, e.g., Leibniz to de Volder, January 21, 1704 (G II 261–262); June 30, 1704 (G II 267); January 19, 1706 (G II 282). See also, e.g., Leibniz to Lady Masham, July 10, 1705 (G III 367).
15. See also Leibniz to Des Bosses, July 31, 1709, LDB 140. I am citing the letters to Des Bosses only in the excellent new edition of Brandon Look and Donald Rutherford, which includes both a new Latin text and a translation, as well as much material not in the version in G II. I would like especially to thank them for having shared their work with me before the volume was published.
16. See also the letters of February 15, 1612, LDB 224–226; September 20, 1712, LDB 276.
17. See also Des Bosses to Leibniz, June 25, 1707, LDB 90; April 22, 1709, LDB 120; July 30, 1709, LDB 134–136; May 20, 1712, LDB 236–238; December 12, 1712, LDB 286.
18. There is a variant of this in the Des Bosses letters, on which the reality of body rests now on what we might call the Divine dream, on what God is supposed to perceive, but I will not discuss that here. See Leibniz to Des Bosses, February 15, 1712, LDB 230–232. See also Leibniz to Des Bosses, January 24, 1713, LDB 296.
19. Paul Lodge has recently transcribed supplements to Leibniz's last letter to de Volder of January 19, 1706, that present a similar position:
> I do not see what argument could prove that there is anything in extension, bulk, or motion beyond the phenomena, i.e. beyond the perceptions of simple substances.
>
> Arguments, in my opinion, cannot prove the existence of anything besides perceivers and perceptions (if you subtract their common cause), and the things that should be admitted in them. In a perceiver,

these are the transitions from perception to perception, with the same subject remaining; in perceptions, the harmony of perceivers. For the rest, we invent natures of things and wrestle with the chimeras of our minds as if with ghosts.

These will appear in his forthcoming edition of the Leibniz–de Volder letters.
20. See, e.g. Leibniz to Jaquelot, March 22, 1703, G III 457; "Eclaircissements sur les natures plastiques . . ." (Appendix to "Considerations sur les natures plastiques . . . ," May 1705), G IV 550; Leibniz to Des Maizeaux, July 8, 1711, G VII 535; Leibniz to Remond, November 4, 1715, G III 657; etc.
21. Interestingly enough, such a view is *not* found in the companion piece to the *Principles of Nature and Grace*, the *Monadology*, where Leibniz seems to go out of his way to avoid using the term "substance" for anything but monads.
22. In my understanding of this view of Leibniz's I am deeply indebted to the excellent introduction to LDB by the editors and translators, Brandon Look and Donald Rutherford. This introduction builds on a monograph and series of articles by Look, where he develops aspects of Leibniz's idea of a substantial bond. See the monograph, Look 1999, and the articles Look 1998, 2000, 2002, and 2004.
23. Donald Rutherford makes a similar remark in Rutherford 2008: 149, 153–154, though he takes it in a somewhat different direction than I do here. I think that we came upon this insight independently and at roughly the same time, in connection with a symposium on Leibniz and Idealism at the APA Central Division in April 2005.
24. Eisler 1910, 1: 516, lists Christian Wolff's *Psychologia rationalis* (1734) and Alexander Baumgarten's *Metaphysica* (1739) as the earliest appearances of the words "idealism" and "idealist" used in the modern sense. The *Trésor de la langue française* lists Diderot's *Lettre sur les aveugles* (1749) in this connection. It is interesting that in the Wolff, *Psychologia rationalis* § 36, where the term is introduced ("*Idealistae* dicuntur, qui nonnisi idealem corporum in animabus nostris existentiam concedunt . . ."), the sole example he gives of an idealist is not Leibniz, whose metaphysics he knew well, but Berkeley. Leibniz does use the word "idealist" once in his writings, in the "Reponse aux reflexions contenues dans la seconde Edition du Dictionnaire Critique de M. Bayle . . . ," written in 1702, published in 1716. In discussing his pre-established harmony he notes: " . . . what is of value in the theories of Epicurus and of Plato, of the greatest materialists and the greatest idealists, is united here" (G IV 560 (WF 114)). However, here the term seems to point to Plato's doctrine of the forms, as contrasted with Epicurus's materialism, rather than the modern conception of idealism, as Wolff defines it.
25. For reasons somewhat different than mine, Justin Smith has arrived at a similar conclusion. See Smith 2004, where he tries to place the question of Leibniz's supposed realism and idealism about body into the context of later Platonism.

BIBLIOGRAPHY

Adams, R. M. (1994) *Leibniz: Determinist, Theist, Idealist*, New York: Oxford University Press.
Baumgarten, A. (1739) *Metaphysica*, Halle: Hemmerde.
Eisler, R. (1910) *Wörterbuch der philosophischen Begriffe*, 3rd edn, 3 vols, Berlin: Ernst Siegfried Mittler und Sohn.
Diderot, Denis (1749) *Lettre sur les aveugles a l'usage de ceux qui voyent*, London [Paris?]: n.p.

Garber, D. (1992) *Descartes' Metaphysical Physics*, Chicago: University of Chicago Press.
——. (2009) *Leibniz: Body, Substance, Monad*, Oxford: Oxford University Press.
Leibniz, G. W. (1734) *Recueil de diverses pieces sur la philosophie, les mathematiques, l'histoire &c. par M. de Leibniz*, ed. C. Kortholt, Hamburg: Abram Vandenhoeck.
——. (1849–1863) *Leibnizens mathematische Schriften*, ed. C. I. Gerhardt, 7 vols, Berlin: A. Asher.
——. (1875–1890) *Die philosophischen Schriften*, ed. C. I. Gerhardt, 7 vols, Berlin: Weidmann.
——. (1923–) *Sämtliche Schriften und Briefe*, eds. Deutsche Akademie der Wissenschaften zu Berlin, Berlin: Akademie-Verlag.
——(1989) *Philosophical Essays*, eds. and trans. R. Ariew and D. Garber, Indianapolis: Hackett.
——. (1997) *Leibniz's "New system" and Associated Contemporary Texts*, eds. and trans. R. S. Woolhouse and Richard Francks, Oxford: Oxford University Press.
——. (2007) *The Leibniz–Des Bosses Correspondence*, eds. and trans. B. Look and D. Rutherford, New Haven: Yale University Press.
Leibniz, G. W. and Arnauld, A. (1967) *The Leibniz–Arnauld Correspondence*, ed. and trans. H. T. Mason, Manchester: Manchester University Press.
Lodge, P., ed. (2004) *Leibniz and his Correspondents*, Cambridge: Cambridge University Press.
Look, B. (1998) "From the metaphysical union of mind and body to the real union of monads: Leibniz on *Supposita* and *Vincula Substantialia*," *Southern Journal of Philosophy*, 36: 505–529.
——. (1999) *Leibniz and the "Vinculum Substantiale"* (Studia Leibnitiana Sonderheft 30), Stuttgart: F. Steiner.
——. (2000) "Leibniz and the substance of the *Vinculum Substantiale*," *Journal of the History of Philosophy*, 38: 203–220.
——. (2002) "On monadic domination in Leibniz's metaphysics," *British Journal for the History of Philosophy*, 10: 379–399.
——. (2004) "On substance and relations in Leibniz's correspondence with Des Bosses" in ed. P. Lodge 2004, pp. 238–261.
Rutherford, D. (2008) "Leibniz as Idealist," *Oxford Studies in Early Modern Philosophy*, 4: 141–190.
Sleigh, R. C. (1990) *Leibniz and Arnauld: A Commentary on their Correspondence*, New Haven: Yale University Press.
Smith, J. E. H. (2004) "Christian Platonism and the metaphysics of body in Leibniz," *British Journal for the History of Philosophy*, 12: 43–59.
Wolff, C. (1734) *Psychologia rationalis*, Frankfurt and Leipzig: Officina libraria Rengeriana.

10 Leibniz on Void and Matter

Sorin Costreie

The principle that the universe is a plenum has two interesting consequences for Leibniz's conception of matter. First, if there is no void, then everything is full of matter and there are no ultimate atoms of the substantial world, and thus matter should be infinitely divisible. Second, if there is no void, then there is no action-at-a-distance and thus (Newtonian) universal gravitation is not acceptable within the framework of this system. This chapter discusses these two consequences of Leibniz's explicit denial of the existence of void.

THE INFINITE DIVISIBILITY OF MATTER

In order to see what void is, one should first clarify the notion of matter, for void is commonly conceived of as lack of matter: it is empty space. Void/matter is a fundamental dichotomy, and void or vacuum is simply the negation of matter or plenum. The locus classicus of the rejection of the void is Aristotle's *Physics* (Book IV), where, contrary to the Greek atomists, Aristotle explicitly denies the existence of a "void" (a region of space containing no substance), holding instead that the universe is filled with continuous substance. As a result "nature abhors a vacuum" became a foundational doctrine of European metaphysical thinking for centuries. Leibniz inherited this *horror vacui*,[1] and in a sense his position concerning the vacuum evolved in response to this point. But let us see how Leibniz conceives of void and matter.

Leibniz's account of matter did not remain static throughout his adult life, but rather it underwent substantial changes.[2] Yet, in at least one respect his view was consistent: from the very beginning he maintained the well-known thesis that matter is infinitely divisible.[3] Moreover, for Leibniz, matter is not only potentially divisible but actually divided. He also endorsed the idea that actual bodies are discrete objects, and so we come upon the paradoxical view that every finite portion of matter comprises infinitely many finite parts. The division of material things can continue *ad infinitum*, which implies that there are no minimal or ultimate material elements of matter. Just as a linear continuum is not made up of points, there are no substantial atoms in the world either. The actual existence of atoms is impossible because, for Leibniz, there are no two identical items in the world. No atoms, no void. *Prima facie*, this seems to be correct, yet one could think of a world of atoms without interspatial void among them, yet the problem will be to understand how motion is possible in such a frozen universe.

Leibniz's explicit rejection of void is a direct consequence of his principle of plenitude, which, in turn, derives from the principle of perfection or principle of the best: *the actual world is the best possible world*.[4] Moreover, it may be said that this metaphysical optimism is based on his fundamental principle that everything has a cause (*nihil sine ratione*), which in conjunction with the fact that God is the supreme benevolent being and ruler of the universe, it follows rationally that we actually live in the best possible world. Our world comprises the maximal amount of perfection, so it cannot leave room for any possible void spaces. Void spaces cannot also be accepted lest we have two identical items in the world, which again would contradict the principle of plenitude. God is an omnipotent gardener, who fills up every little tiny portion of the universe with matter, with infinitely decomposable matter.

Let us see now what Leibniz means by "infinitely decomposable matter." Johann Bernoulli, in a letter from August 1698, expresses to Leibniz his doubts concerning this puzzling idea in the following way: "You admit that any portion of matter is already actually divided up into an infinite number of parts; and yet you deny that any of these parts can be infinitely small. How is this consistent?" (GM iii 529). Leibniz's reaction was to resort to mathematics:

> I do not think it follows from (the infinite division of matter) that there exists any infinitely small portion of matter. Still less do I admit that it follows that there is any absolutely minimum portion of matter ... Let us suppose that in a line its 1/2, 1/4, 1/8, 1/16, 1/32, etc., are actually assigned, and that all the terms of this series actually exist. You infer from this that there also exists an infinitieth term. I, on the other hand, think that nothing follows from this other than that there actually exists any assignable finite fraction, however small you please.
>
> (GM III 536)

Thus, in order better to grasp Leibniz's thought, it is quite important to see how this infinite division of matter is captured by a mathematical model. In general, for Leibniz, mathematics provides us an ultimate model for the intelligibility of the world. In the present case the model is offered by the sequence 1/2, 1/4, 1/8, 1/16, 1/32 ... The idea behind this sequence is that the process of infinite division is infinite precisely in the sense that the sequence has no last term. The physical counterpart of this process is the infinite decomposition of matter. There is no last or least unit of matter, just because, as in the case of natural numbers, given any certain number in this series, we can always find a smaller number, which is in fact its successor in this convergent sequence. It is worth noting that here, as in the case of natural numbers, we can only speak of a *syncategorematic*[5] actual infinite, and one cannot argue in favor of a categorematic actual infinite. The latter may imply the actual existence of infinitesimal magnitudes, an idea strongly rejected by Leibniz.

With regard to this mathematical model, Samuel Levey (1999: 143–144) provides us an interesting analysis, suggesting two alternative ways of interpreting the infinite division of matter. One is called the "diminishing pennies" model:

> Imagine a glass jar halfway full of metal coins, say copper pennies. Resting squarely on the bottom of the jar, one coin is uniquely the largest of them all: this is the *alpha penny*. Otherwise like an ordinary penny, the alpha penny is one-half inch thick. Also inside the jar there is a penny, just one, which, like all the pennies in the jar, is ordinary in its width and circumference but which is exactly half as thick as the alpha penny; this second penny measures one-quarter inch in thickness. . . . And so on, ad infinitum. . . . There is no 'omega penny', so to speak, of only infinitesimal thickness, to be found anywhere in the jar.

The other model is called the "divided block" model and runs as follows:

> Here's a block of stone one cubic foot in volume and regular in its dimensions: it measures one foot high by one foot long by one foot wide. On close inspection it is observed that the block is neatly divided down the middle by a hairline fissure into two equal slabs, each one foot high by one foot long by only half a foot wide. . . . And so on ad infinitum. . . . Each and every slab is seen, in fact, to give way to two smaller slabs, without end.

Both models consist of two different ways of representing the infinite division of matter. These models in combination with Leibniz's own model lead Levey to the conclusion that Leibniz seems to be contradictory here, in the sense that Leibniz's model does not fit his own ontological considerations concerning the infinite division of matter. I will show that Levey's analysis is incorrect, for it does not cover all Leibnizian suggestions concerning this issue, especially the innumerability of parts of each portion of matter. I shall turn to this point later. Consider now Leibniz's own model (folding tunic), which runs as follows in *Pacidus to Philalethes*:

> [T]he division of the continuum must not be considered to be like the division of sands into grains, but like that of a sheet of paper or tunic into folds. And so although there occur some folds smaller than others infinite in number, a body is never thereby dissolved into points or minima. On the contrary, every liquid (i.e., all matter) has some tenacity so that although it is torn into parts, not all the parts of the parts are so torn in their turn; instead at any time they merely take shape, and are transformed; and yet in this way there is no dissolution all the way down into points. . . . It is just as if we suppose a tunic to be scored with folds multiplied to infinity in such a way that there is no fold so small that it is not subdivided by a new fold. . . . And the tunic cannot be said to be resolved all the way down into points; instead, although some folds are smaller than others to infinity, bodies are always extended and points never become parts, but always remain mere extrema.
>
> (A.VI.iii.78: 555; ALC 185–187)

When Leibniz speaks here about parts that are infinite in number, he has in mind the syncategorematic sense of the term *infinite*. He conceives of matter in this way because here he deals with finite material bodies, even though they can be infinitely divisible. Moreover, they are actually infinitely divided in infinitely many finite parts. Leibniz thinks that there are no ultimate material constituents of the substantial world, there are no material atoms or points. Each portion of matter, regardless of how small it is, has extension and is further divided to infinity. Last but not least, Leibniz is rather ambiguous here in discussing matter in a way in which it seems that it is continuous, whereas, on the contrary, he is very clear in other places that substantial bodies *qua* aggregates are discrete. So, in what sense can matter be seen as an infinitely decomposable continuum, and yet as actually made up of finite bodies?

THE LABYRINTH OF THE CONTINUUM

For Leibniz, there are two important labyrinths of knowledge given by the nature of the continuum and of freedom.

> There are two famous labyrinths where our reason very often goes astray: one concerns the great question of the Free and the necessary, above all in the production and the origin of Evil; the other consists in the discussion of continuity and of the indivisibles which appear to be the elements thereof, and where the consideration of the infinite must enter in.
>
> (T 53)

With regard to the labyrinth of the infinite decomposition of the continuum, let us recall what Leibniz explicitly states in a letter to Foucher:

I am so in favour of the *actual infinite* that instead of admitting that Nature abhors it, as is commonly said, I hold that Nature makes frequent use of it everywhere, in order to show more effectively the perfections of its Author. Thus I believe that there is no part of matter which is not, I do not say divisible, but *actually divided*; and consequently the least particle ought to be considered as a world full of an infinity of different creatures.

(G I 416)

This conception of the actual infinite division of matter raises at least three questions:

1. How can a discrete finite piece of matter be actually divided into an infinity of smaller parts?
2. Moreover, does this imply that we should acknowledge the existence of some infinitely small yet indivisible parts of matter? (Bernoulli's question)
3. What (mathematical) model(s) could be used to "visualize" this infinite divisibility of matter into smaller parts?

1. Normally, only the continuum is regarded as being (potentially) divisible into an infinity of smaller parts. But for Leibniz matter is *discrete* and thus we need a new account of how this infinite divisibility works in the case of a discrete piece of matter, and, moreover, how this infinity of parts is not merely potential but *actual*. The basic idea of the Leibnizian solution is that "discreteness" is opposed to "continuity" but not to "contiguousness," and thus in any piece of matter reside an infinity of smaller *contiguous* parts. Each part is individuated by a different motion, and they do not constitute a real whole, unless they are an aggregate united by a mind: "matter alone is explicable by a multiplicity without continuity. . . . matter is a discrete entity, not a continuous one; it is only *contiguous*, and is united by motion or by a mind of some sort" (A.VI.iii.60: 474; ALC 47, my italics).

2. With regard to the second question, recall that Leibniz's answer is quite explicit: *infinitesimals* are only useful fictions. They constitute a very precious tool in applying mathematics to physical reality, but cannot be found as such in the real world; they are only a *façon de parler*. It should be added here that for Leibniz there are two types of objects: mathematical objects, which are ideal and *continuous*, and where it is said that the whole is prior to its parts, and physical objects, which are real and *discrete*, and where it is said that the parts are prior to the whole. To confuse them constitutes a great mistake, which quite often has led the human minds into the *labyrinth of the continuum*[6]: "it is the confusion of the ideal and the actual that has embroiled everything and produced the labyrinth of the composition of the continuum" (G ii 98; ALC xxiv).[7] However, for Leibniz, there is a bridge between mathematics and

physics, and this is given by mathematics, more precisely by *infinitesimal analysis*[8]:

> The infinitesimal calculus is useful with respect to the application of mathematics to physics; however, that is not how I claim to account for the nature of things. For I consider infinitesimal quantities to be useful fictions.
>
> (G VI 629; AG 230)

This is an important point, which concerns the relation between mathematics and physics, and raises the following question: how can Leibniz accept infinitesimal magnitudes in physics while at the same time rejecting them in mathematics? Daniel Garber has recently tackled this issue, providing a fine analysis. His answer to this question is:

> Leibniz's point in distinguishing mathematics from the physical world is not to reject the mathematical representation of physical magnitudes, but simply for us to understand what is going on when we mathematize nature, and what the role of mathematics is in the understanding of nature. His opponents are the Cartesians who have tried to make nature mathematical in a literal sense, to make the physical world over into a physical instantiation of mathematical concepts. He wants to reject this, but in doing so, he wants only to restore mathematics to its proper place in the enterprise, and not to reject it altogether. In this way we can embrace the mathematical representation of dead force in terms of infinitesimals, without having to say that there are real infinitesimals in nature.
>
> (Garber 2008: 305–306)

Mathematics is a very important tool for investigating nature, and provides us a model of the physical world, yet it is just a model, an abstract picture that cannot be found as such in the real world. I shall come back to this point later, at the end of the chapter.

3. One mathematical model of this infinite divisibility is offered by the arithmetical model when considering the sum of the infinite convergent series of 1/2, 1/4, 1/8, 1/16, 1/32 . . . The sum is obviously 1 and thus we can mathematically figure out how a finite portion of matter (for example a one-meter bar) can be divided into two equal parts, and one of them will be further divided into another two parts and so forth. It is clear that each resulting part is extended, and in our case it will be, in fact, greater than 0, but smaller than 1. Since there is no last or least term, there will be no final indivisible part of matter, simply because, as in the case of natural numbers, given any certain number in this series, we can find immediately a smaller number, which will be in fact its successor in this convergent series.

The basic idea is that there are no ultimate material units of the corporeal world. Each portion of matter, however small it is, has extension and can be further divided to infinity. I do not contest that, judging so, Leibniz's model is closer to Levey's divided block model and thus it might raise some problems with respect to the actual divisibility of matter, but I claim that there is another mathematical model that can be found in Leibniz's writings, and that has a remarkable geometrical nature and better fits his intuitions. I call it the *worlds within worlds* model. The basic metaphysical idea behind this model is that in every piece of matter there is, in fact, a huge density of matter, and thus any tiny piece of matter comprises a whole world of other material things. It is like saying that in any atom we can find a whole universe with many galaxies, with a infinity of stars and planets. The geometrical representation of this model is based on representing various circles circumscribed in one initial circle,[9] which constitutes our original piece of matter. This geometrical model seems to be a more appropriate representation of Leibniz's metaphysical idea of the actually infinite divisibility of matter than the other arithmetic model of the "folded tunic." The former makes explicit the connection with two important points: matter as an aggregate composed of various *contiguous* parts and matter as actually *folded*. By "folds" Leibniz meant that we cannot infinitely divide matter up to extensionless points (which by their very nature will be indivisible), but into other actually divisible portions of matter that have extension and that are like very tiny *strings of matter*. In this sense each piece of matter can be conceived of as a shrinking manifold of contiguous parts, where each part in turn is constituted by another shrinking manifold and so on.

> [T]here could be, indeed, there have to be, *worlds* not inferior in beauty and variety to ours in the smallest motes of dust, indeed, in tiny atoms. And (what could be considered even more amazing) nothing prevents animals from being transported to such worlds by dying, for I think that death is nothing but the contraction of an animal, just as generation is nothing but its *unfolding*.
> (GM III 553; AG 169, my italics)

> [T]here is no place so small that we cannot imagine a smaller sphere to exist in it. If we suppose this to be so, there will be no assignable place that is empty. And yet the world will be a plenum, from which it is understood that an unassignable quantity is something.
> (A.VI.iii.76:525; ALC 61)

> All worlds are *contiguous*. It follows from the same thing that everywhere there is a world.
> (A.VI.ii.N45; ALC 346)

"Worlds within worlds" reads geometrically as follows: infinitely many circles inside bigger circles, plus other infinitely many circles that "reside" in the space among the contiguous circles, such that the result is that in fact there is no empty space, since everywhere we can imagine some tiny little circle that fits within the other circles. Certainly, this idea could easily be extended to the case of three-dimensional bodies, where circles become spheres.

> [I]t suffices for a body to be integral on the surface; for inside it is again composed of infinitely many globes, and new worlds can be contained in it without end.
> (ALC xlvi)

> [A]ny atom will be of infinite species, like a sort of world, and there will be *worlds within worlds to infinity.*
> (A.VI.ii:241; ALC xliv)

This interpretation finds textual support in Leibniz's writing, even though, to the best of my knowledge, it was never explicitly propounded by Leibniz as a "visual" solution for the infinite decomposition of matter. In this respect, one should consider the following passages:

> There is no portion of matter that is not actually divided into further parts, so there is no body so small that there is not a world of infinitary creatures in it.
> (A.VI.iii.78:566; ALC 209)

> In any grain of sand whatever there is not just a world, but even an infinity of worlds.
> (A.VI.iii.78:566; ALC 211)

> There is an infinite number of creatures in the smallest particle of matter, because of the actual division of the continuum to infinity.
> (T 195)

> [T]here is no place so small that we cannot imagine a smaller sphere to exist in it. If we suppose this to be so, there will be no assignable place that is empty. And yet the world will be a plenum, from which it is understood that an unassignable quantity is something. . . . [F]or given the plenitude of the world, it is necessary that there exist some globules smaller than others to infinity.
> (A.VI.iii.76:525; ALC 61)

> Everything is a plenum.... Matter is actually divided into infinite parts. There are infinitely many creatures in any body whatever. All bodies cohere to one another.... there are no Atoms.
>
> (A.VI.ii.N42; ALC 344)

> We may suppose the whole space of the world to be filled with globes, touching each other only at a point, so that the interstices cut out between them (*intercepta*) are vacua. But all integral bodies, i.e. those lacking an interposed vacuum, are uniform, i.e. spherical.
>
> (A.VI.ii.N44; ALC 344)

Why is this model better than the alternatives proposed by Levey? I think that one important advantage of this model is that it contains an important feature of actual infinity, namely *uncountability*.[10] Theoretically, in Levey's models, even if we go on *ad infinitum*, we are able to count how many parts a piece of matter has, since we can easily find a one-to-one correspondence with the series of positive integers. But Leibniz says explicitly in various places, and at different times of his life that:

> I concede the infinite multiplicity of terms, but this multiplicity does not constitute a number or a single whole. It signifies only that *there are more terms than can be designated by a number*.
>
> (GM III 575; ALC lxii)[11]

> Any part whatever of a body is a body, by the very definition of a body. So bodies are actually infinite, i.e. *more bodies can be found than there are unities in any given number.*
>
> (A.VI.iv.266; ALC lxiii, my italics)

But perhaps the most explicit passages in this sense could be seen the following couple:

> Now the cause of these things is that the Creator left nothing sterile, but instead fashioned a kind of world of innumerable creatures (*innumerabi[ium] creaturarum*) in any particle of matter whatever, however minute.
>
> (A.iv.301; ALC 291)

> [T]here are no atoms, but every part again has parts actually divided from each other and excited by different motions, or what follows from this, every body however small has actually infinite parts, and in every grain of powder there is a world on *innumerable* creatures.
>
> (A.VI.iv.312: 1623; ALC lxvi)

The underlying idea of these passages is that the density of matter comprised in every tiny piece of matter is far beyond any "countable" density we could ever empirically discover. Precisely as Cantor would explain hundreds years later, there is a difference between the cardinality of the set of natural numbers and that of real numbers. Leibniz's metaphysical model of worlds within worlds resembles the geometrical counterpart of the arithmetical model of the continuum of real numbers. Besides this specific characteristic that we cannot count the parts of a body, there is yet another general Leibnizian consideration that entitles us to think that the worlds within worlds model is better than any arithmetical model:

> [O]nly *Geometry* can provide a thread for the Labyrinth of the Composition of the Continuum, of maximum and minimum, and the unassignable and the infinite, and no one will arrive at a truly solid metaphysics who has not passed through that labyrinth.
> (A.VI.iii.54: 449; ALC xxiii)

I would like to add that this model of worlds within worlds, conjoined with the principle of sufficient reason, may be used by Leibniz to explain the existence of a necessary cause, which lies outside the series of particular contingent truths. Graham Priest has given a detailed discussion of this point.[12] However, he does not have any particular geometrical Leibnizian model in mind, but his whole discussion is based on the acknowledgment in Leibniz's system of a higher density of matter, which is neither decomposable in minimal points, nor countable. In Priest's terms this is called "the possibility of infinitude after infinitude."

We may now join Levey in wondering whether Leibniz was aware of the tension between accepting an actual infinity of things and rejecting the existence of an infinite number. I believe that the answer is positive. He acknowledged the difficulty in doing so, and this seems to undermine Levey's interpretation, since he considers that the source of the problem is exactly the conflict between the acceptance of the actual infinite in metaphysics and the rejection of it in mathematics. Leibniz says explicitly that:

> It is perfectly correct to say that *there is an infinity of things*, i.e. that there are always more of them than one can specify, but it is easy to demonstrate that there is no infinite number, nor any infinite line, nor other infinite quantity, if these are taken to be genuine wholes.
> (G VII 468)

Therefore, the rejection of the void has one important consequence of Leibniz's account of matter: its infinite decomposability. There is just matter, and in this ontological scheme there is no room for void, which may be conceived of only as a mere possibility.

LEIBNIZ'S REJECTION OF NEWTON'S GRAVITY

Another important consequence of Leibniz's rejection of the void is his disapproval of the action-at-a-distance. It is well known that one point on which Leibniz is at odds with Newton is the issue of gravity, and I think that this is a direct consequence of his rejection of a vacuum. In *Anti-barbarus physicus* (G vii 343; AG 318) Leibniz says that the defenders of gravity "are forced by this view of the essential attraction of matter to defend the vacuum, since the attraction of everything for everything else would be pointless if everything were full. But in the true philosophy, the vacuum is rejected for other reasons." As I said at the very beginning, for Leibniz, there cannot be any void in the actual world, because it is the best possible world and thus the space of that vacuum could have been filled with another finite substance, thus increasing the perfection of the world even further, but since it is the best possible world, its perfection is maximal so there is no vacuum in our universe. Moreover, the existence of void would violate Leibniz's identity of indiscernibles, for in a vacuum it is possible to have two identical portions of space. But this again violates the maximal perfection of the actual world and ultimately the great principle of sufficient reason. There is no reason for a benevolent God to create two identical things in the most perfect world, because a world with more diversity, that has no identical items in it, would have been more perfect; therefore the perfection of the best possible world requires a plenum. Thus, this metaphysical rejection of vacuum has at least another two important consequences in his physics: the rejection of universal attraction between bodies and the rejection of absolute space, for both would require the existence of void.

Leibniz is suspicious of Newton's universal gravitation because it is contrary to his metaphysical principles: it presupposes actions among substances and requires the constant intervention of God into the physical world, which is seen by Leibniz as being contrary to the divine perfection of the creator, who acts as a perfect designer who sets up from the very beginning a world that does not require any further intervention and adjustment. Action-at-a-distance is in Leibniz's eyes miraculous. In a letter to Lady Masham, he speaks quite clearly with respect to this matter:

> [I]t is no more comprehensible to say that a body acts at a distance with no means or intermediary than it is to say that substances as different as the soul and the body operate on each other immediately; for there is a greater gap between their two natures than between any two places. So the communication between these two so heterogeneous substances can only be brought about by a miracle, as can the immediate communication between two distant bodies; and to try to attribute it to I know not that what influence of the on the other is to disguise the miracle with meaningless words.
>
> (G III 354)

But this raises the following question: why did Leibniz end up rejecting only universal gravitation, namely action-at-a-distance, while still accepting mind–body connection? Why is it easier to account for the mind–body interaction than for its body–body counterpart?

Newton affirms in his *Optics*[13] that God is a "powerful ever-living agent, who being in all places, is more able by his will to move the bodies within his boundless uniform *sensorium*, and thereby to form and reform the parts of the universe, than we are by our will to move the parts of our bodies." This makes Leibniz react by attacking the idea of (absolute) space as *sensorium Dei*, along with the idea that thus God is continuously (and miraculously!) intervening in the world. As a response to this, in his fourth letter to Leibniz, Clarke takes a step back and reacts in the following way:

> That one body should attract another without any intermediate means, is indeed not a miracle, but a contradiction; for 'tis supposing something to act where it is not. But the means by which two bodies attract each other, may be invisible and intangible, and of a different nature from mechanism; and yet, acting regularly and constantly, may well be called natural.
>
> (G VII 388; LC 53)

So, just because it is constant, for Clarke, and implicitly for Newton, God's intervention is natural and thus not miraculous, whereas for Leibniz, precisely because it is God's intervention, it would still be a miracle. Yet, being continuous would contradict the nature of true miracles, which are sporadic interventions of God, in order to supply the order of grace and not that of nature. Leibniz states clearly that:

> According to my opinion, the same force and vigour remains always in the world, and only passes from one matter to another, agreeably to the laws of nature, and the beautiful pre-established order. And I hold, that when God works miracles, he does not do it in order to supply the wants of nature, but those of grace. Whoever thinks otherwise, must need have a very mean notion of the wisdom of God.
>
> (LC 12)

This is an important point of the discussion, because in a sense it pertains to the maintenance role played by God. For Newton, the denial of God's continuous direct intervention in the natural world would, in fact, somehow mean the deprivation of God's rights as the supreme ruler, who, as a benevolent monarch, should constantly intervene into the affairs of his kingdom. For Leibniz, something is miraculous whether it surpasses the power of created beings or transcends human knowledge. There are *epistemological* miracles, in the sense that the laws of the actual world cannot be comprehended by any created mind, and *ontological* miracles, namely those by which the order of the world is truly violated by a supernatural power. But, God in his supreme benevolence, would not make physical laws epistemologically miraculous, for we have the power to discover them in the order of nature studied purely geometrically. This brings us to another related issue, namely, to the connection between physics and metaphysics. Leibniz manifests his dissatisfaction with dealing only with the geometrical order of the world:

> There was a time when I believed that all the phenomena of motion could be explained on purely geometrical principles, assuming no metaphysical propositions, and that the laws of impact depend only on the composition of motions. But, through more profound meditation, I discovered that this is impossible, and I learned a truth higher than all mechanics, namely, that everything in nature can indeed be explained mechanically, but that the principles of mechanics themselves depend on metaphysical and, in a sense, moral principles, that is, on the contemplation of the most perfectly effectual [*operans*], efficient and final cause, namely, *God*, and cannot in any way be deduced from the blind composition of motions. And thus, I learned that is impossible for there to be nothing in the world except matter and its variations, as the Epicureans held.
> (G vii 280; AG 245–246)

Gregory Brown sees here the initiation of the construction of a methodological wall between physics and metaphysics, or, in his terms, the seeds of "the methodological Apartheid between Physics and Metaphysics" (Brown 2007: 150). I would not go so far in characterizing the relation between the two disciplines, or, better terms, between two levels of investigating the universe. I would rather join Garber in saying that: "Though Leibniz was clearly an adherent of the new mechanical philosophy by the late 1660s, there is no reason to believe that he thought that his brand of mechanism was in any way inconsistent with an adherence to Aristotelian philosophy. . . . All and all, it seems best to view Leibniz in the context not of the radical mechanists, but of the renovators or reformers, of the seventeenth century thinkers who were attracted to the new mechanical philosophy while at the same time thinking that it could be reconciled with the old Aristotelian physics" (Garber 1995: 273). Perhaps a Hegelian would say in this context that Aristotle's substantial forms were the thesis, Descartes' mechanical philosophy of nature the antithesis, and, eventually, Leibniz's mature conception of the world (of monads) the synthesis. However, I agree with Brown that there is a certain methodological wall between Leibniz and Newton, since the latter says the following:

> I have not as yet been able to deduce from phenomena the reason for these properties of gravity, and I do not feign hypotheses. For whatever

is not deduced from the phenomena must be called a hypothesis; and hypotheses, whether metaphysical or physical, or based on occult qualities, or mechanical, have no place in experimental philosophy. In this experimental philosophy, propositions are deduced from the phenomena and are made general by induction. The impenetrability, mobility, and impetus of bodies, and the laws of motion and the law of gravity have been found by this method. And it is enough that gravity really exists and acts according to the laws that we have set forth and is sufficient to explain all the motions of the heavenly bodies and of our sea.
(Newton 1999: 943)

Newton defines "force" solely on the basis of its measurable effects without any reference to its metaphysical origin. Thus, he constructed a systematic account that yields dynamical predictions by defining "force" in terms of acceleration. On the other hand, Leibniz thought that physics alone could not provide the ultimate knowledge of the world, unless it is based on the more fundamental metaphysical problems. This is an important point of divergence between Leibniz and Newton, and this disagreement concerns both the way they regard the universe as a whole and the way in which we may attain scientific knowledge in such a universe. Yet, as Brown himself observes, the wall between the two is in a sense unidirectional, for Leibniz also appreciates the great benefits of the experimental method in science, since, in a letter to A. Conti, he confesses:

I . . . admire very much the physico-mathematical thoughts of M. Newton, and the public would be greatly in your debt, sir, if you could persuade this capable man to give us his conjunctures in physics to this point. I strongly approve his method of deducing from phenomena what can be deduced from them without supposing anything, though this would sometimes be only to deduce conjectural consequences. However, when the data are not sufficient, it is permitted (as is done in deciphering) to conceive hypotheses, and if they succeed, we accept them provisionally, until some new experiences bring us new data, and what Bacon calls a crucial experiment for deciding between the hypotheses. . . . I am very much in favor of the experimental philosophy, but M. Newton errs greatly when he supposes that all matter is heavy (or that each part of matter attracts every other part), which experience does not establish, as M. Huygens has already very rightly determined. The gravitating matter cannot itself have this heaviness of which it is the cause, and M. Newton does not furnish any experience or sufficient reason for the void and atoms, or for universal mutual attraction. And since it is not known perfectly and in detail how gravity is produced, or elastic force, or magnetic force etc., we do not on that account have reason to appeal to scholastic occult qualities or to miracles.
(Newton 1959–1977, 6: 252–253)

The problem of induction seems to mark a major difference between Newton and Leibniz. For the former it is the true and sole path to attain solid knowledge about the empirical world, whereas for Leibniz, neither the senses nor induction provide us secure knowledge about the world. In a letter to Queen Sophie, he points out that knowledge of the universal truth of science is possible not because of something outside us, but rather because of something inside, i.e. our innate knowledge:

> [T]here is a light which is born in us. For since the senses and induction can never teach us truths that are fully universal or absolutely necessary, but only what is and what is found in particular examples, and since we nonetheless know the universal and necessary truths of the science—in this we are privileged above the beasts—it follows that we have drawn these truths in part from what is within us.
> (L 551)

Later on, Leibniz becomes clear about the metaphysical differences between these two kinds of interaction: body–body, through action at a distance, and mind–body, through pre-established harmony:

> There is no comparison between the action of one body on another and the influence of the soul on the body. There is immediate contact between bodies, and we understand how that can be, and how, since there is no penetration, their coming together must alter their movement in some way. But we see no such consequences with the soul and the body: these two do not touch, and do not interfere with one another in an immediate way which we can understand and deduce from their natures. All they can do is to be in agreement, and to depend on one another by a metaphysical influence, so to speak, in virtue of the soul's ideas; and that is contained in God's plans, which are in conformity with them. They are related through the mediation of God, not by a continual interruption of the laws of the one because of its relation to the other, by a harmony pre-established in their natures *une fois pour toutes*.
> (G vi 570)

Summing up, I think that an interesting conclusion we may draw from this analysis is that, in order to understand Leibniz's notion of matter and its place in his mature philosophical system,[14] one should acknowledge the existence of three levels of analysis: the metaphysical level of monads as the ultimately real level of reality; the physical level of phenomena, which are grounded on monads and their states; and the mathematical level of ideal entities, which *qua ens rationis* (numbers, infinitesimals, space, time ...) are "useful fictions" that help us better understand the gap between the first two levels of analysis of the universe.[15] Thus, the connections and the foundational role played by each become apparent in this scheme. Metaphysics plays a foundational role for physics, whereas mathematics is a useful tool, an ideal conceptual net that the mind throws upon the world in order to capture the laws of its phenomena. This tripartite scheme of analysis is based on an essential ontological tripartition. There are three kinds of units in the world. First, there are true units, which are metaphysical and lie at the basis of everything; they are the monads, and are both real and exact. Second, there are also exact but not real units, which are mathematical and they are simple possibilities; they make the world intelligible for us. Thirdly, there are real but inexact units, which are physical; they are the constitutive elements of phenomena. *Matter* is constituted at this level as a well-founded phenomenon.

> But *atoms of matter* are contrary to reason. Furthermore, they are still composed of parts, since the invincible attachment of one part to another (if we can reasonably conceive or assume this) does not eliminate diversity of those parts. There are only *atoms of substance*, that is, real unities absolutely destitute of parts, which are the source of actions, the first absolute principles of the composition of things, and, as it were, the final elements in the analysis of substantial things. We could call them *metaphysical points*: they have *something vital*, a kind of *perception*, and *mathematical points* are the *points of view* from which they express the universe. But when corporeal substances are contracted, all their organs together constitute only a *physical point* relative to us. Thus physical points are indivisible only in appearance; mathematical points are exact, but they are merely modalities. Only metaphysical points or points of substance (constituted by forms or souls) are exact and real, and without them there would be nothing real, since without true unities there would be no multitude.
>
> (AG 142)

George Gale (1970: 237) goes further and sees the following tripartition:

> This three-level system constitutes the basic essentials of Leibniz's complete well-founded philosophical system. The system is a *system* simply because each level is in strict correspondence to the other, that is all phenomenal levels can be shown to be well-founded (mathematically derivative) from the metaphysical level. Stated in its simplest terms Leibniz's view is that observable phenomena (derivative forces) are well-founded upon the explanatory phenomena of physics (primitive forces), and the explanatory phenomena of physics are well-founded upon the entities of metaphysics (Monads and their perception).

The difference between Gale's scheme and mine is that he sees at the level of forces an ontological divide between the physical and the phenomenal world. I find this problematic simply because for Leibniz "[p]hysics deals with the matter of things and the unique affection resulting from the combination of matter with the other causes, that is to say, with motion. . . . *Matter in itself is devoid of motion. Mind is the principle of all motion*" (L 99). So, the science of dynamics plus the study of collisions, both dealing with derivative forces, are physical sciences. But a discussion *in extenso* of this point is beyond the scope of this chapter and, in fact, does not affect the general point presented here.[15]

ACKNOWLEDGMENTS

Many thanks for comments and suggestions to Daniel Garber, John Bell, Peter Anstey, Norma Goethe, Ilie Parvu, and to the members of the Romanian Early Modern Research Group: Dana Jalobeanu, Sorana Corneanu, Mihnea Dobre, and Lucian Petrescu. Thanks also to Petruta Naidut, who provided me some useful linguistic hints. This paper was supported within The Knowledge Based Society Project supported by the Sectorial Operational Program Human Resources Development (SOP HRD), financed by the European Social Fund and by the Romanian Government under the contract no. POSDRU ID 56815.

NOTES

1. A.VI.iii.585: "there is no vacuum, whether interspersed or great, since it is possible for all things to be filled."
2. For an excellent account of this development see Garber 2009.
3. With regard to these issues (void, infinitely decomposable matter, gravity . . .) Leibniz did not substantially change his view over the years; thus the present chapter focuses on the systematic connection among these points, without paying much attention to their historical development in Leibniz's thinking. This does not mean that I adhere to Russell's point that there is just one static Leibnizian philosophical system; it means simply that, as Garber has nicely formulated, "below the surface flux of hesitations, changes of mind, there may be some very deep commitments, commitments about organism and mentality, unity and substantiality, force and activity, and the basic intelligibility of the world, commitments he came upon and worked out in a gradual way, commitments that shape and limit what appear to us as swings in his mature thought" (Garber 1985: 74). Leibniz's rejection of void is one of his deep commitments.
4. For more details concerning Leibniz's basic principles and the constitutive relations among them see Mates 1986: Ch. 9 and Rescher 1979: Ch. 3.
5. "It is perfectly correct to say that *there is an infinity of things*, i.e. that there are always more of them than one can specify, but it is easy to demonstrate that there is no infinite number, nor any infinite line, nor other infinite quantity, if these are taken to be genuine wholes. The Scholastics were taking that view, or should have been doing so, when they allowed a 'syncategorematic' infinite, as they called it, but not a 'categorematic' one. The true infinite, strictly speaking, is only in the *absolute*, which precedes all composition and is not formed by the addition of parts" (G vii 468). In fact, Leibniz's conception of the infinite is even more complicated, since

he also considers a third alternative, that of a "hypercategorematic infinite": "There is a *syncategorematic* infinite, or a passive power having parts, namely, the possibility of further progression in dividing, multiplying, subtracting, and adding. There is also a *hypercategorematic* infinite, or a potestative infinite, an active power having parts, as it were, eminently, not formally or actually. This infinite is God himself. But there is no *categorematic* infinite, or one actually having infinite parts formally" (G ii 314f). More on this distinction can be found in Moore 1990: 51: "Roughly: to use 'infinite' *categorematically* is to say that there is something which has a property that surpasses any finite measure; to use it *syncategorematically* is to say that, given any finite measure, there is something which has a property that surpasses it." Arthur (2001) draws an interesting distinction, which seems to carry an important insight. For Arthur, Leibniz is consistent in his claims concerning infinity, and should be understood as endorsing *an infinite that is actual but syncategorematic*. Arthur (2001: 46) says: "Leibniz has distinguished a perfectly middle ground between Cantor's actual infinite and Aristotle's potential infinite. Claims that it is some sort of unsound halfway house between the two are in error: it is quite coherent to say that there are actually infinitely many things without accepting Cantor's theory of the transfinite." Carlin (1997: 10) points out that "it is important to realize, of course, that Leibniz's denial of the categorematic infinite is not a denial of the actual infinite."

6. "A new and unexpected light finally arose in a quarter where I least hoped for it—namely, out of mathematical considerations of the nature of the infinite. There are two labyrinths of the human mind: one concerns the composition of the continuum, and the other the nature of freedom, and both spring from the same source—the infinite" (T 107).
7. See also Russell 1992: 245: "In actuals, single terms are prior to aggregates, in ideals the whole is prior to the part. The neglect of this consideration has brought forth the labyrinth of the continuum" (G ii 379).
8. See also Russell 1992: 233: "The infinitesimal analysis has given us the means of allying Geometry with Physics" (G v 370).
9. Circles, or, of course, in real three-dimensional space, spheres.
10. There is "a kind of a world of innumerable creatures in any particle of matter whatever, however minute" (A.VI.iv.301: 1511; ALC 281).
11. Letter to Bernoulli in 1699, my italics.
12. Priest 1995, sect. 2.8. *Leibniz's repair:* 36–38.
13. Query 23 of the 1706 edition; Query 31 of the 1717 edition, LC 181.
14. See the previous chapter for more details.
15. This picture is not new and it was endorsed by Winterbourne 1982 and Hartz and Cover 1988, and partially by Rescher 1979 and Mates 1986.

BIBLIOGRAPHY

Arthur, R. (1998) "Infinite aggregates and phenomenal wholes: Leibniz's theory of substance as a solution to the continuum problem," *Leibniz Society Review*, 8: 25–45.
———. (1999) "Infinite number and the World Soul: in defence of Carlin and Leibniz," *Leibniz Society Review*, 9: 105–116.
———. (2001) "Leibniz and Cantor on the actual infinite" in ed. H. Poser 2001, pp. 41–46.
Brown, G. (1998) "Who's afraid of infinite numbers?: Leibniz and the World Soul," *Leibniz Society Review*, 8: 113–125.

———. (2000) "Leibniz on wholes, unities and infinite numbers," *Leibniz Society Review*, 10: 21–51.

———. (2005) "Leibniz's mathematical argument against a soul of the world," *British Journal for the History of Philosophy*, 13: 449–488.

———. (2007) "Is the logic in London different from the logic in Hanover?: some methodological issues in Leibniz's dispute with the Newtonians over the cause of gravity" in eds. P. Phemister and S. Brown 2007, pp. 145–162.

Carlin, L. (1997) "Infinite accumulations and pantheistic implications: Leibniz and the *Anima Mundi*," *Leibniz Society Review*, 7: 1–24.

Gale, G. (1970) "The physical theory of Leibniz" in ed. R. S. Woolhouse 1994, pp. 227–239.

Garber, D. (1985) "Leibniz and the foundations of physics: the middle years" in eds. K. Okruhlik and J. R. Brown 1985, pp. 27–130.

———. (1995) "Leibniz: physics and philosophy" in ed. N. Jolley 1985, pp. 270–352.

———. (2008) "Dead force, infinitesimals, and the mathematicization of nature," in eds. U. Goldenbaum and D. Jesseph 2008, pp. 281–306.

———. (2009) *Leibniz: Body, Substance, Monad*, Oxford: Oxford University Press.

Gennaro, R. and Huenemann, C., eds. (1999) *New Essays on the Rationalists*, Oxford: Oxford University Press.

Goldenbaum, U. and Jesseph, D., eds. (2008) *Infinitesimal Differences: Controversies between Leibniz and his Contemporaries*, Berlin: Walter de Gruyter.

Hartz, G. A. and Cover, J. A. (1988) "Space and time in the Leibnizian metaphysics" in ed. R. S. Woolhouse 1994, pp. 76–103.

Jolley, N., ed. (1995) *The Cambridge Companion to Leibniz*, Cambridge: Cambridge University Press.

Leibniz, G. W. (1923–) *Sämtliche Schriften und Briefe*, eds. Deutsche Akademie der Wissenschaften zu Berlin, Berlin: Akademie-Verlag.

———. (1849–1863) *Leibnizens mathematische Schriften*, 7 vols, ed. C. I. Gerhardt, Berlin: A. Asher.

———. (1875–1890) *Die philosophischen Schriften*, 7 vols, ed. C. I. Gerhardt, Berlin: Weidmann.

———. (1952) *Theodicy. Essays on the Goodness of God, the Freedom of Man, and the Origin of Evil*, ed. A. Farrar; trans. E. M. Huggard, La Salle, Il; Open Court.

———. (1956) *The Leibniz–Clarke Correspondence*, ed. H. G. Alexander; trans. H. T. Mason, Manchester and New York: Manchester University Press and Barnes & Noble.

———. (1969) *G. W. Leibniz: Philosophical Papers and Letters*, 2nd edn, ed. and trans. L. E. Loemker, Dordrecht: D. Reidel.

———. (1989) *Philosophical Essays*, eds. and trans. R. Ariew and D. Garber, Indianapolis: Hackett.

———. (2001) *The Labyrinth of the Continuum: Writings on the Continuum Problem, 1672–1686*, Trans., ed. and comm. R. Arthur, New Haven: Yale University Press.

Levey, S. (1998) "Leibniz on mathematics and the actual infinite division of matter," *Philosophical Review*, 107: 49–96.

———. (1999) "Leibniz's constructivism and infinitely folded matter" in eds. R. Gennaro and C. Huenemann 1999, pp. 134–163.

Mates, B. (1986) *The Philosophy of Leibniz: Metaphysics and Language*, Oxford: Oxford University Press.

Moore, A. (1990) *The Infinite*, London: Routledge.

Newton, Sir I. (1959–1977) *The Correspondence of Isaac Newton*, 7 vols, eds. H. W. Turnbull, J. F. Scott, A. R. Hall and L. Tilling, Cambridge: Cambridge University Press.

———. (1999) *The Principia: Mathematical Principles of Natural Philosophy*, eds. and trans. I. B. Cohen and A. Whitman, Berkeley: University of California Press; 1st edn 1687.

Okruhlik, K. and Brown, J. R., eds. (1985) *The Natural Philosophy of Leibniz*, Dordrecht: Reidel.

Papineau, D. (1977) "The vis viva controversy: do meanings matter?" in ed. R. S. Woolhouse 1994, pp. 198–216.

Phemister, P. and Brown, S., eds. (2007) *Leibniz and the English-Speaking World*, Dordrecht: Springer.

Poser, H., ed. (2001) *Proceedings of the VII International Leibniz Congress*, Hanover: G. W. Leibniz–Gesellschaft.

Priest, G. (1995) *Beyond the Limits of Thought*, Oxford: Clarendon Press.

Rescher, N. (1979) *Leibniz: An Introduction to his Philosophy*, Oxford: Blackwell.

Russell, B. (1992) *The Philosophy of Leibniz*, London: Routledge.

Winterbourne, A. T. (1982) "On the metaphysics of Leibnizian space and time" in ed. R. S. Woolhouse 1994, pp. 62–75.

Woolhouse, R. S., ed. (1994) *Gottfried Wilhelm Leibniz: Critical Assessments*, vol. 3: "Philosophy of science, logic and language," London: Routledge.

11 Hume on the Distinction between Primary and Secondary Qualities

Jani Hakkarainen

INTRODUCTION

George Berkeley is well known for his immaterialist philosophy and metaphysics; the good Bishop is one of the main critics of the "new" corpuscularian and Cartesian philosophy of body of the seventeenth century. Hume's attack on substance is standard material in introductions to the history of philosophy as well as his skepticism concerning the external world, or as he himself calls it, skepticism against or with regard to the senses. It is also common knowledge that Hume uses one of Berkeley's many skeptical arguments against matter in his writings. This is the argument at the end of Part 1, Section 12 of *An Enquiry concerning Human Understanding* (1748)[1] (EHU 12.15–16) and the main argument of the Section *Of the modern philosophy* in *A Treatise of Human Nature* (1739)[2] (*Treatise* 1.4.4.5–16). It is therefore surprising that this Berkeleyan skeptical argument is too much ignored in the literature (Hakkarainen 2007: xvii–xviii). The argument is apparently premised on the distinction between primary and secondary qualities. Accordingly, Hume's understanding and view of this distinction has not received enough attention in the literature either. In general, his attitude to the "new" philosophy and metaphysics of body is not widely studied.

This chapter is part of a larger study of Hume's metaphysics of body and his attitude to Metaphysical Realism,[3] which is part of one of the hottest debates of Hume scholarship in the last decades, namely, "the New Hume Debate" on whether or not he is a Causal Realist (Richman and Read 2007). One of the key issues in my project is Hume's attitude to the Berkeleyan skeptical argument that is apparently premised on the primary/secondary quality distinction. If that argument really is Hume's own argument, he cannot be a Metaphysical Realist and his philosophy and metaphysics of body must be non-Realist and nonmaterialist. In that case, Hume would join the ranks of the critics of "new" material substance, such as Berkeley, though with a more moderate or less dogmatic attitude. According to the conclusion of this argument of Hume, the notion of material substance is vacuous, empty of content (but the notion is not inconsistent as Berkeley argues). In this sense, Hume would thus think that material substance is incomprehensible to the human understanding.

How we should interpret Hume's metaphysics of body hangs, therefore, on his view of this skeptical argument. Indeed, we should talk about arguments here as the arguments in the first *Enquiry* (EHU 12.15–16) and the *Treatise* (1.4.4.6–15) differ slightly. Following Hume, I will call the first *Enquiry* version of this reasoning "the second profound argument against the senses" and the form that it takes in the *Treatise*, *Treatise* 1.4.4.6–15.

At bottom, Hume's endorsement of these arguments depends on his view of the primary/secondary quality distinction because they are premised on this distinction, or more precisely, on what Hume thinks is fundamental to the distinction. As every other major premise clearly represents his view, it follows that if Hume holds the fundamental principle of this distinction, it is difficult to deny his commitment to the conclusion of the arguments. Thus, in that case his metaphysics of body would be nonmaterialist and his philosophy non-Realist in general.

Accordingly, some Hume scholars who embrace a Realist interpretation have questioned Hume's endorsement of the primary/secondary quality distinction. One of Don Garrett's interpretative strategies in his Realist and naturalist interpretation of Hume[4] is to challenge Hume's assent to the distinction between primary and secondary qualities (Garrett 1997: 218–220). Annette Baier, Donald W. Livingston, and William Edward Morris have denied Hume's endorsement of the entire second profound argument and *Treatise* 1.4.4.6–15 including his endorsement of the primary/secondary quality distinction (Baier 1991: 21 and 107, Livingston 1984: 24 and 9ff., and Morris 2000: 96–102 and 106).[5]

The second part of this chapter discusses Hume's attitude to what he thinks is the fundamental principle of the primary/secondary quality distinction. However, before that we must consider Hume's understanding of this distinction. In the first part of the chapter, therefore I suggest an insight that Hume has into the primary/secondary quality distinction. In order to see this, we must make the distinction between *secondary qualities* and *proper sensibles*. Secondary qualities are powers in Real bodies, while proper sensibles are properties that are the immediate objects of each of our five senses. My first thesis is that Hume has an insight into the heart of most of "new philosophy" when he claims that according to it, proper sensibles are not Real properties of material substance and Real bodies.[6] Accordingly, this is my focus on the distinction between primary and secondary qualities. I will call it "the Proper Sensibles Principle" (PSP).

In the second part of the chapter, I defend the interpretation—mainly against Garrett's doubts—that the Proper Sensibles Principle is a rational tenet in Hume's view and he thus endorses it. Its rationality means that the PSP has a firm foundation in inductive-causal reasoning. My argument has both a positive and a negative part. First I discuss passages from Hume's writings that support his endorsement of the PSP and its rationality. In the negative part, I reply to four challenges to Hume's assent to the PSP, three of which are Garrett's. Part of this section consists in discussing Hume's argument for the PSP at *Treatise* 1.4.4.3–4, which leads to perceiving a gap in the argument observed by Louis Loeb.

HUME'S INSIGHT INTO THE PRIMARY/SECONDARY QUALITY DISTINCTION

Hume seems to be guilty of a serious misunderstanding when he speaks about "secondary" and "sensible qualities" interchangeably in EHU 12.15–16. This apparent misunderstanding is especially manifest in the beginning of the second profound argument against the senses in which Hume writes as follows:

> It is universally allowed by modern enquirers, that all the sensible qualities of objects, such as hard, soft, hot, cold, white, black, &c. are merely secondary, and exist not in the objects themselves, but are perceptions of the mind, without any external archetype or model, which they represent.
> (EHU 12.15)

In this passage, Hume seems to suggest that according to every "new philosopher"—in opposition to medieval or ancient predecessors—secondary qualities are nothing in Real objects. It is plain that Hume's thesis is incorrect in the trivial sense that there were immaterialist new philosophers before him and in his time who did not commit themselves to any claims about the properties of material substance and Real bodies—Berkeley, for example. Let us focus, however, on materialist or substance-dualist new philosophers. Hume's contention appears to be mistaken because not every materialist or dualist new philosopher denied the existence of secondary qualities in material substance and Real bodies. We must only consider Locke, who does not deny the Reality of secondary qualities. Instead, he affirms that they are something in Real bodies, i.e., powers:

> 2*dly*, Such *Qualities*, which in truth are nothing in the Objects themselves, but Powers to produce various Sensations in us by their *primary Qualities*, *i.e.* by the Bulk, Figure, Texture, and Motion of their insensible parts, as Colours, Sounds, Tasts, *etc*. These I call *secondary Qualities*.
> (*Essay* II. viii. 10)[7]

Is Hume then on the wrong track right from the start in the first *Enquiry* when he argues against the comprehensibility of material substance and Real bodies? Is he really such a careless reader of Locke that he misses these passages? Or is he merely misrepresenting Locke on purpose?

I think this is a hasty conclusion for two reasons. First, I would not attribute so gross misunderstanding of Locke to Hume because Locke says so many times that secondary qualities are powers. Secondly, it misses Hume's point. In order to see that, let us consider the corresponding section of the *Treatise* (1.4.4).

When we read this section carefully, we come to realise that Hume uses the term "secondary quality" only twice and never claims that they are nothing in material substance and Real bodies (*Treatise* 1.4.4.9 and 11).[8] Instead, his focus is on what Aristotle called "proper sensibles" (*idia aisthêta*) (*De Anim.* 2.6)[9]: "colours, sounds, tastes, smells, heat and cold" (*Treatise* 1.4.4.3). Although Hume did not use this term, it was common parlance in philosophy of his time and he thinks these qualities were perceived by one sense only (*Treatise* 1.2.3.15 and 1.4.5.11–13). So it is justified to use this term in describing Hume's view.

One of the premises of this section then is that it is proper sensibles that are nothing in material substance and Real bodies. Here are notable examples:

> The fundamental principle of that philosophy is the opinion concerning colours, sounds, tastes, smells, heat and cold; which it asserts to be nothing but impressions in the mind, deriv'd from the operation of external objects, and without any resemblance to the qualities of the objects.
> (*Treatise* 1.4.4.3)

> [U]pon the removal of sounds, colours, heat, cold, and other sensible qualities, from the rank of continu'd independent existences, we are reduc'd merely to what are called primary qualities, as the only *real* ones, of which we have any adequate notion.
> (ibid.: 5)

> If colours, sounds, tastes, and smells be merely perceptions, nothing we can conceive is possest of a real, continu'd, and independent existence.
> (ibid.: 6)

> When we reason from cause and effect, we conclude, that neither colour, sound, taste, nor smell have a continu'd and independent existence. When we exclude these sensible qualities there remains nothing in the universe, which has such an existence.
> (ibid.: 15)

As these quotations show, the point of the primary/secondary quality distinction—what Hume calls "the fundamental principle" of the modern philosophy—is, at least in the view of the *Treatise*, the Proper Sensibles

Principle (PSP). This principle states that proper sensibles are not Real properties of material substance and Real bodies.

It is also to be observed that in the second and fourth quotations Hume speaks about "sensible qualities" instead of "secondary qualities."[10] This clue is important for two reasons. First, it suggests that for understanding Hume's insight we must make the distinction between proper sensibles and secondary qualities. In this context, Hume's use of "sensible qualities" is to refer to proper sensibles.[11] Secondly, it helps to realise that Hume's point is the same in the first *Enquiry* as in the *Treatise*. His focus is rather on proper sensibles than on secondary qualities. In order to see that, let us consider the relevant passages in EHU 12.15–16.

In the second sentence of EHU 12.15, which I have already quoted, Hume first speaks about sensible qualities. Then he gives examples that belong exactly to proper, tactile, and visual sensibles: "hard, soft, hot, cold, white, black." Hume claims that these "are merely secondary," which means that they "exist not in the objects themselves." In the fourth sentence, Hume also speaks about sensible qualities when he writes that "all the qualities, perceived by the senses, be in the mind, not in the object." In the light of these two sentences, it is justified to read the term "sensible qualities" in the next paragraph (16) as meaning proper sensibles. The principle about which Hume is talking there is thus the Proper Sensibles Principle:

> if it be a principle of reason that all sensible [proper sensible] qualities are in the mind, not in the object.
>
> (EHU 12.16)

Furthermore, Hume ends the previous paragraph by arguing that the putative abstract ideas of primary qualities are incomprehensible. In this argument, he concentrates on extension and claims that it cannot be conceived of without tactile or visual properties, hardness, temperature, and colors, which are proper sensibles.

I take these passages to support the interpretation that Hume's point also in the first *Enquiry* is the PSP. At this point I do not want to take any stance on his other views of the primary/secondary quality distinction—as, for example, his view of the status of the putative primary qualities (whether they are Real). My thesis is restricted to denying the Reality of proper sensibles although the PSP implies that the perceptions of proper sensibles cannot resemble anything in material substance and Real bodies.

Showing that Hume's insight into the heart of new philosophy is really an insight would naturally require a very extensive argument discussing all the main figures of the new philosophy. As there is no room for that here, I can only observe that it is a widely accepted view that almost all materialist or dualist new philosophers believed in the PSP. Actually, it was one of the distinguishing features of new philosophy that material substance and Real bodies were cleared out of what Hume took to be the proper sensible properties.[12] In this respect, matter and bodies do not resemble our sense-perceptions. Although Hume abstracts here very much from the diverse new philosophies of body and metaphysics, this seems to be a matter of fact—in spite of the fact that there was no commonly accepted terminology.[13] Take, for example, Descartes and Malebranche. They do not use the terms "primary qualities" and "secondary qualities" but make the distinction between the "essence" and "modifications" of material substance and its "sensible" or "sense qualities" (Malebranche 1997 1.10.1: 49, Elucidation 6: 569–570, and 573–574; Cottingham 1993: 149; and Garber 1992: 292–298).[14] Indeed, this very terminological choice does not undermine Hume's insight but actually supports it. In their terms, Descartes and Malebranche sharply distinguish the Real properties of material substance from proper sensibles.

However, things are not so simple as the standard view suggests. One of the leading figures of new philosophy, Locke, takes solidity as both a proper sensible (touch) and a primary quality (*Essay* II. iv. 1 and II. viii. 9). This is a clear counter-example to Hume's insight casting a shadow on the universality of its extension. Nevertheless, I do not think that it can disprove Hume's claim; his insight seems to be a correct *general* description of the core of the materialist and dualist new philosophy. Besides, he can and does argue against Locke in this respect: for Hume, solidity is not a proper sensible but a complex common sensible (touch and sight) (*Treatise* 1.4.4.9 and 12–14).

HUME'S VIEW OF THE PROPER SENSIBLES PRINCIPLE AND ITS RATIONALITY

Textual Evidence

The first potential evidence for Hume holding the rationality of the PSP is the point that the first *Enquiry* and the *Treatise* treat it rather as a rational than as an unreflective principle. The PSP is classified as a philosophical tenet instead of an everyday, natural or "vulgar" principle. It is attributed to "modern enquirers" or "modern philosophy."

It must be admitted, however, that this is only potential evidence. Hume may still think that this principle is not a proper rational tenet. This possibility would be that the argument for it does not really justify it; this argument is not a good argument. Hence we need more evidence for the interpretation that Hume endorses the PSP because it is rational.

As I have observed above, Hume claims that the PSP is a principle "universally allowed by modern enquirers" (EHU 12.15). Notwithstanding its historical correctness, this quotation yields evidence for Hume's endorsement

Hume on the Distinction between Primary and Secondary Qualities   241

of the principle itself if he includes himself among these modern enquirers. He asserts that the PSP is universally believed by them and if he belongs to these modern enquirers, this quotation is a statement of his view.

In Hume's time, as in ours, "modern" was understood as the opposite to "ancient" (OED: modern, 2.a; Hume 1987: 245) and Hume most certainly does not think he is one of the ancients. Therefore, it seems to be a reasonable assumption that he identifies himself with these modern enquirers and assents to the PSP. However, this may not be the whole story. "Modern enquirers" can refer to a subgroup of the modern philosophers and we cannot be sure that Hume thinks he belongs to that group. Consider, for instance, the distinction between Cartesian and Newtonian physics. Both are "modern" but still there are deep differences between them. Thus, this quote cannot settle the problem we are addressing now.

The next natural move is to look at the entire first *Enquiry* and to determine whether it can provide evidence of Hume's attitude to the rationality of the PSP or his relationship to the modern enquirers. Unfortunately, there is no explicit evidence for this apart from a withdrawn footnote (EHU 1.14.n.1; 1748–1750 editions). That note concerns morality and is relevant here. However, in addition to being omitted from the final editions, it must be discussed in connection with similar passages from Hume's other texts. Let us therefore consider what the *Treatise*, his essays and other texts can say about Hume's attitude to the PSP and its rationality.

My thesis is that when all the relevant passages in Hume's texts are taken into consideration, the evidence supports his endorsement of the rationality of the PSP, and the principle itself, rather than his suspension or denial of it, and this, despite the fact that Hume is more explicit of his assent in the *Treatise* than in the later texts. We are, therefore, justified in claiming that Hume endorses the PSP as a rational tenet. This does not imply, however, that Hume's assent to the PSP is unqualified—or that the evidence is decisive. I will discuss the certainty of his assent below.

To begin with, in 1762 Hume commented on Reid's manuscript of *An Inquiry into the Human Mind on the Principles of Common Sense* (1764). This happened through the hands of their common friend, Hugh Blair (1718–1800).[15] A part of this short letter is relevant for the present purposes:

> The Author supposes, that the Vulgar do not believe the sensible Qualities of Heat, Smell, Sound, & probably Colour to be really in the Bodies, but only their Causes or something capable of producing them in the Mind. But this is imagining the Vulgar to be Philosophers & Corpuscularians from their Infancy. You know what pains it cost Malebranche & Locke to establish that Principle. There are but obscure Traces of it among the Antients viz in the Epicurean School. The Peripatetics maintaind opposite Principles. And indeed

> Philosophy scarce ever advances a greater Paradox in the Eyes of the People, than when it affirms that Snow is neither cold nor white: Fire hot nor red.
> (Hume to Hugh Blair, July 4, 1762; Hume 1986: 416)

Hume begins this passage by stating the PSP. His point is to criticise Reid for contending that this principle represents our everyday view. His argument is that from the everyday point of view the PSP is paradoxical and it was Malebranche and Locke who proved the tenet. So it is rather a rational than a "vulgar" or natural principle and a paradox only from the everyday perspective. This letter suggests then that Hume takes the PSP as a proven, rational tenet.

It must be granted, however, that this letter in itself cannot constitute decisive evidence for Hume holding the rationality of the PSP. Further and more substantial support can be found from Hume's famous essay *Of the Standard of Taste*. It was published in 1757 and as such it is an important document of Hume's mature views. When he establishes one of the key claims of the essay that "there are certain general principles of [aesthetic] approbation or blame" in human nature, Hume writes as follows:

> If, in the sound state of the organ, there be an entire or a considerable uniformity of sentiment among men, we may thence derive an idea of the perfect beauty; in like manner as the appearance of objects in day-light, to the eye of a man in health, is denominated their true and real colour, even while *colour is allowed to be merely a phantasm of the senses.*
> (Hume 1987: 234; emphasis added)

This passage is typical of Hume in the way it compares aesthetic and moral beauty with proper sensibles; colors at this point. Here he argues that we can determine "the perfect beauty" in a similar way to our deciding the "true and real colour" of things. Regarding beauty, the situation becomes more complex later in the essay. Still Hume does challenge the view that some works of art, like those by Virgil, are uniformly felt beautiful (ibid.: 242–243). So here Hume is reporting his own view of beauty, which is only qualified later. As he tries to convince his reader by comparing it with colors, it is most likely that he is reporting his own position with regard to the latter as well—this is obviously the effect he wants to generate here. That position involves the proposition that color is "merely a phantasm of the senses," which is a special case of the PSP. This passage constitutes therefore evidence for the interpretation that Hume endorses not only the rationality of the PSP but also the principle itself.

It should also be taken into account that the PSP specific to (physical) taste appears in a positive light at the beginning of the essay. When Hume discusses "a species of philosophy, which cuts off all hopes of success in" attempting "to seek a Standard of Taste," he writes that to "seek in the real beauty, or real deformity, is as fruitless an enquiry, as to pretend to ascertain the real sweet or real bitter" (Hume 1987, 229–230). In this article, there is no room to establish that this "species of philosophy," interpreted in a certain way at least, is the starting point of Hume's argument in the essay (which is not based on false premises) and it represents his own view. But if we presume that, and that is what we ought to do in my view, this quote also supports the interpretation that the PSP (of taste) is genuinely a Humean principle. This account is only confirmed by what Hume writes later in the essay:

> Though *it be certain that beauty and deformity, more than sweet and bitter, are not qualities in objects*, but belong entirely to the sentiment, internal or external, it must be allowed, that there are certain qualities in objects which are fitted by nature to produce those particular feelings.
> (Hume 1987: 235; emphasis added)

It is possible to trace this comparison of beauty with proper sensibles back to the *Treatise* through Hume's other essay, *The Sceptic*. It was published about one year later than the third Book of the *Treatise*, in January 1742 (Miller 1987: xii–xiii). The following footnote, more theoretical nature than the essay itself, was attached by Hume to the passage where he claims that the beauty of Virgil's *Aeneid* (c. 30 BC.; unfinished at his death) "lies not in the poem, but in the sentiment or taste of the reader." The footnote is so relevant for the purposes of this paper that it is justified to quote it in its entirety:

> Were I not afraid of appearing too philosophical, I should remind my reader of that famous doctrine, supposed to be fully proved in modern times, "That tastes and colours, and all other sensible qualities, lie not in the bodies, but merely in the senses." The case is the same with beauty and deformity, virtue and vice. This doctrine, however, takes off no more from the reality of the latter qualities, than from that of the former; nor need it give any umbrage either to critics or moralists. Though colours were allowed to lie only in the eye, would dyers or painters ever be less regarded or esteemed? There is sufficient uniformity in the senses and feelings of mankind, to make all these qualities the objects of art and reasoning, and to have the greatest influence on life and manners. And as it is certain, that *the discovery* above-mentioned *in natural philosophy*, makes no alteration on action and conduct; why should *a like discovery in moral philosophy* make any alteration?
> (Hume 1987: 166, n.3; emphases added)

This passage is similar to those in *Of the Standard of Taste*; note for instance that Virgil is there again to illustrate the point. The difference between these passages lies in the exposition because this time Hume is more explicit about his own views. First he claims that the PSP is "supposed to be fully proved in modern times." This reminds us of his letter to Blair where Hume says that Malebranche and Locke took pains to establish it. Although "supposed" in itself may cause us to doubt whether Hume really thinks that the PSP is "fully proved," the letter to Blair and the beginning of the quotation read together strongly suggest that Hume takes the PSP to be decisively proven. This interpretation is corroborated by how Hume continues the footnote. He identifies the PSP with the non-Reality of beauty and deformity: "the case is the same with beauty and deformity." He also refers to both by the phrase "This doctrine." As the non-Reality of beauty and deformity is, without doubt, Hume's own position, in the light of this passage, the PSP must be as well. They are the same and it is not possible to hold one without the other. Moreover, at the end of the quotation, Hume writes that the PSP is a "discovery ... in natural philosophy" in the same manner as the non-Reality of beauty and deformity is in moral philosophy.

This claim of the discovery occurs also in a closely resembling passage in the third Book of the *Treatise* before *The Sceptic*. In that passage as well, Hume compares (moral) beauty and deformity with the PSP:

> Vice and virtue, therefore, may be compar'd to sounds, colours, heat and cold, which, according to modern philosophy, are not qualities in objects, but perceptions in the mind: And this discovery in morals, *like that other in physics, is to be regarded as a considerable advancement* of the speculative sciences; tho', like that too, it has little or no influence on practice.
>
> (*Treatise* 3.1.1.26; emphasis added)

This passage is important because now Hume does not speak about the supposed proof or mere discovery. He writes that the PSP is a discovery that *must* be taken as an important theoretical improvement, especially in physics. So this yields compelling evidence for Hume's endorsement of the rationality of the PSP, and indeed, of the principle itself.

With these passages in hand, it is time to discuss the footnote that was omitted from the later editions of the first *Enquiry*. The relevant part of the footnote is the following:

> But a late Philosopher [Hutcheson] has taught us, by the most convincing Arguments, that Morality is nothing in the abstract Nature of Things, but is entirely relative to the Sentiment or mental Taste of each particular Being; *in the same Manner as the Distinction of sweet and bitter, hot and cold, arise from the particular Feeling of each Sense and Organ.*
>
> (EHU 1.14.n.1; 1748–1750 editions; emphasis added)

In this passage, Hume makes the same comparison between beauty and proper sensibles that he does in the writings he kept on publishing. Thus,

Hume did not drop the footnote due to a change of mind. The natural explanation is that he cut it off after 1750 because of the publication of the second *Enquiry*, *An Enquiry concerning the Principles of Morals*, in the following year. It follows from this that the footnote is relevant to Hume's considered view of the PSP. Indeed, it provides firm support for my claim that Hume endorses the rationality of the PSP. At the end of the passage, Hume writes without reservations that the distinctions between sweet and bitter, hot and cold "arise" due to the perceiver. They are therefore "relative to the Sentiment or mental Taste of each particular Being." Thus, tastes and felt temperatures are not Real in their ontological status and Hume is stating a restricted PSP here. The nice thing about the passage is that Hume explicitly commits to the restricted PSP; he is not merely describing Hutcheson's views.

In the *Treatise*, the soundest evidence can be found from 1.4.4 and Hume's reference back to it in the Conclusion of Book 1 (1.4.7).[16] Hume begins *Treatise* 1.4.4 by bringing forward the "fundamental principle of that philosophy [modern]," the PSP, and comments its foundation as follows:

> Upon examination, I find only one of the reasons commonly produc'd for this opinion to be satisfactory, *viz.* that deriv'd from the variations of those impressions, even while the external object, to all appearance, continues the same.
>
> (*Treatise* 1.4.4.3)

In the next paragraph, he continues by claiming that: "The conclusion drawn from them [sense variations], is likewise *as satisfactory as can possibly be imagin'd.*" (*Treatise* 1.4.4.4; emphasis added). This conclusion is partly based on the principle that "from like effects we presume like causes" (ibid.).

These passages show that in the *Treatise* Hume takes the PSP as an inductive-causally rational principle. First, he claims that there are "satisfactory" reasons for it. In other words, Hume thinks that there is a convincing argument that justifies the principle. Secondly, the conclusion of the argument, i.e., the PSP, is not only convincing but convincing to the highest imaginable and possible degree. In the third place, the argument for the PSP works on the premise that is Hume's fourth "rule" of causal reasoning or inductive inference in the *Treatise* (1.3.15.6). This argument is then a just inductive inference from Hume's point of view. This is confirmed by the last paragraph of *Treatise* 1.4.4 and Hume's reference back to it in 1.4.7:

> "When we reason from cause and effect, we conclude, that neither colour, sound, taste, nor smell have a continu'd and independent existence." (*Treatise* 1.4.4.15); "nor is it possible for us to reason justly and regularly from causes and effects, and at the same time believe the continu'd existence of matter."
>
> (*Treatise* 1.4.7.4).

Besides the above-discussed passages, there is only one in Hume's texts that is relevant: EHU 12.16. Peter Millican has suggested that this paragraph may be critical of the PSP because Hume might be advancing a *modus tollens* rather than a *modus ponens* there (Millican 2002b: 465). In this passage, Hume says that if the PSP is rational, the belief in "external existence," i.e., in the existence of Real entities, is contrary to reason. Millican's proposal is that Hume's point might be to challenge the antecedent by the absurdity of the consequent. If this passage is read out of its wider context in Hume's corpus, Millican's proposal may have some claims to be a justified reading of it.

Nonetheless, the overall evidence points in the opposite direction, towards the reading that Hume embraces the antecedent, i.e., the rationality of the PSP. There are good reasons to think that in *Of the Standard of Taste* the PSP appears as a Humean tenet. In *The Sceptic*, Hume explicitly identifies it with his doctrine that beauty is not a Real property, which is repeated in the Hutcheson footnote to EHU 1 in the early editions. *The Sceptic* and the third Book of the *Treatise* treat the PSP also as a significant improvement in natural philosophy. *The Sceptic* and Hume's letter to Blair, read together, take it as a proven principle. In most of these passages, the PSP is not restricted to nonspatial properties of sounds, smells, and tastes (*Treatise* 1.4.5.9–16) but it is extended equally to visual and tactile qualities. The bulk of the quotations are also from the writings Hume kept on publishing. Perhaps the clearest evidence is, however, from *Treatise* 1.4.4 and 1.4.7 where Hume explicitly says that the PSP is an inductively well-grounded tenet.

It may be objected to this that the strongest expressions of the endorsement of the PSP and its rationality are in the *Treatise* and the later texts are deliberately more cautious. This is correct *prima facie*, but it only reflects the general change in Hume's rhetoric. The *Treatise* is written mostly in the first person singular. In the later works, Hume uses more rhetorical devices such as the third person singular, the passive, the verbs "to allow" and "to suppose," the conditional, and the dialogue form that create the impression that the writer detaches himself from the text. The point is, however, that he does so many times when he is, without doubt, putting forward his own views.[17] The impression of caution in the passages under discussion from the later texts cannot therefore undermine their evidential value. Especially in the case of such a difficult and important interpretative question as Hume's attitude to the PSP, we ought not to be led too much by first appearances but must read the passages carefully and compare them with each other.

To sum up, the textual evidence we have of Hume's attitude to the PSP and its rationality provides more support for the reading that Hume takes the PSP as a rational tenet than that he does not. Although this evidence is not decisive, especially in the point of the certainty of Hume's belief in the PSP, I think it is sufficient for the interpretation that

Hume embraces the Proper Sensibles Principle. Its rationality commands his assent.

Challenges and Answers

Notwithstanding the fact that this interpretation has a firm textual basis, four questions can be raised concerning its correctness. The first challenge comes from the direction of non-Realist reading of Hume, the wider argument of which this chapter forms a part. If Hume does not believe in the existence of material substances and Real bodies (non-Realism), how is it possible for him to endorse the PSP that makes a statement about them? This is analogous to the contention of Bricke, Wright, and Donald Baxter that Hume's corresponding argument for the mind-dependency of perceptions in EHU 12.9 and *Treatise* 1.4.2.50 presupposes Metaphysical Realism[18] (Bricke 1980: 20; Wright 1983: 86; Baxter 2008: 14).

This question may be answered by observing that it is completely coherent to formulate the PSP conditionally. This formulation would be that *if* there are both Real material substance and Real bodies, they do not have proper sensibles as their Real properties; the proponent of this principle does not have to believe that there are material substance and Real bodies. Therefore there is no obstacle for a non-Realist to maintain the principle conditionally.[19]

The second critical point is made by Garrett (1997: 218)—actually it is an objection to Hume's assent to the PSP. He quotes Hume from *Treatise* 1.4.4.6 where Hume says that "many objections might be made to this system." Garrett thinks that Hume is here also referring to the PSP as part of the modern "system" of natural philosophy. The PSP would also be a target of *Treatise* 1.4.4.6–15 and a deep shadow is cast on his endorsement of it, not to speak of its rationality.[20]

My reply to Garrett is to reflect on the structure of *Treatise* 1.4.4 and the logic of Hume's arguments there. After relating this section to the previous ("Of the antient philosophy") in the two first paragraphs, the discussion of modern philosophy begins with the PSP and the argument for it in §§3–4. As we have seen, in these paragraphs, Hume's characterizations of the PSP and the argument for it are straightforwardly and strongly positive. The fifth paragraph starts with the statement that "all the other doctrines of that [modern] philosophy seem to follow by an easy consequence" from the PSP. I think that much weight should be put on the qualification "seem" here. It is at this point where Hume's tone becomes more reserved and later sharply critical of modern philosophy. It also shows that Hume takes the PSP to be independent from "the other doctrines of" the modern philosophy.[21] These "other doctrines" are a tenet that may be called "the Primary Qualities Principle" (PQP) and principles of explanation, which are not so relevant for my present argument. The Primary Qualities Principle states that the only comprehensible properties that material substance and Real bodies have are primary qualities:

> [W]e are reduc'd merely to what are call'd primary qualities, as the only *real* ones, of which we have any adequate notion.
> (*Treatise* 1.4.4.5)[23]

Hume's argument against modern philosophy covers the rest of the section. He begins it boldly:

> I believe many objections might be made to this system: But at present I shall confine myself to one, which is in my opinion very decisive.
> (*Treatise* 1.4.4.6)

Garrett's worry stems from this passage because Hume appears to say that he argues against the PSP as well when his target is the system of modern philosophy. However, if we consider what the actual target of Hume's arguments is, we come to realise that it is not at least directly the PSP but the PQP. The conclusion of Hume's arguments is the negation of the PQP. He argues to the result that the notion of material substance and Real body as having only primary qualities is incomprehensible. The PQP is thus false since it states that they are the only comprehensible properties. Moreover, the arguments against the PQP are clearly based on the PSP. Hume appears to argue merely against the PQP (and thus the explanatory principles since according to them, primary qualities do the explanatory work). The PSP is left untouched and Hume has no problem endorsing it just like the beginning of the section strongly suggests.

It is possible, however, that things are not so simple. *Treatise* 1.4.4.6–15 may form a *modus tollens* against the entire system of modern philosophy including the PSP. The supposed *modus tollens* argument would be that this system has a false or absurd consequence. First, I should say that I do not believe that Hume takes *Treatise* 1.4.4.6–15 as a *modus tollens*. My reason for this is that *Treatise* 1.4.4.6–15 is premised not only on the PSP but also on the principles he undoubtedly endorses. For instance, he uses his metaphysics of space established in Part 2, Book 1. However, in order to disprove the *modus tollens* reading, let us suppose for the sake of the argument that *Treatise* 1.4.4.6–15 indeed forms a *modus tollens*.

Hume evidently thinks that if there is a fundamental principle in the system of modern philosophy, it is the PSP. All other principles "seem" to follow from it. This undermines the readings that *Treatise* 1.4.4.6–15 forms a *modus tollens* against the system of modern philosophy as the *conjunction* of the PSP and the PQP, which would prove that the PSP is false. Even if Hume thinks that PSP → PQP is false (that it merely seems to follow), he clearly does not think that the system is their conjunction.

Hume on the Distinction between Primary and Secondary Qualities 249

Thus, *Treatise* 1.4.4.6–15 cannot be a *modus tollens* against the PSP in this sense. The other exhaustive possibility[23] is that the system has the form PSP → PQP.[24] *Treatise* 1.4.4.6–15 would be then a *modus tollens* against this implication. On this assumption, the PSP must be true because PSP → PQP is false (it is disproved by a *reductio ad absurdum*), which fits well with Hume's qualification "seem to follow."

All things considered, it is not a reasonable reading of *Treatise* 1.4.4.6–15 that Hume's "decisive objection" is aimed at the PSP. Accordingly, Garrett's challenge cannot undermine the interpretation that Hume embraces the PSP and its rationality.

Garrett's additional challenge is his claim that Hume "pointedly refrains from endorsing" the PSP (Garrett 1997: 220). His basis for this assertion is that

> Hume himself [. . .] does not ever assert the truth of the modern philosophers' conclusion about the unreality of secondary qualities. Instead, he restricts himself reporting it as their conclusion.
> (Garrett 1997: 218).

It is correct that Hume never explicitly writes in the first person, "I, David Hume, maintain the PSP," or in any corresponding manner. Still I think Garrett's conclusion ought not to be drawn from this fact. I have shown above that there is sufficient textual evidence for Hume assenting to the PSP and its rationality in his writings from different periods, most of which he had republished. The weakness in Garrett's position is that he does not take account of the other texts than the first book of the *Treatise*. He does not advance an argument for denying the evidential value of the passages in *Treatise* 3, *The Sceptic*, the 1748–1750 editions of the first *Enquiry*, and *Of the Standard of Taste*. However, the most remarkable omission in his discussion is that he does not quote or discuss the passage in *Treatise* 1.4.4 where Hume indeed "asserts the truth of the modern philosophers' conclusion about "the unreality of secondary qualities." Hume explicitly claims that the PSP is satisfactory of *the highest possible degree* because it is the conclusion of the argument in *Treatise* 1.4.3.3–4. Garrett clearly omits and downplays important evidence against his view in this respect.

The fourth challenge is rather casting doubt on Hume's endorsement of the PSP than arguing that he refuses to believe it. It is also from Garrett's book and consists in the assessment of the argument for the principle in *Treatise* 1.4.4.3–4 (Garrett 1997: 218–220).

In replying to the fourth question, I shall first outline the argument in both Hume's and my own words. After that, it is possible to assess its justness. In the second part, I first locate what I consider to be the real gap in the argument, which is observed by Loeb. From that discussion, I proceed to discuss Garrett's doubts about the justness of the argument. It leads us

250   Jani Hakkarainen

to judge his doubts and finally to conclude about the certainty of Hume's endorsement of the PSP.[25]

The argument outlined using Hume's own words is the following:

(1) the variations of those impressions, even while the external object, to all appearance, continues the same (*Treatise* 1.4.4.3)
   These variations depend upon several circumstances (ibid.)
   Upon the different situations of our health: A man in a malady feels a disagreeable taste in meats, which before pleas'd him the most (ibid.)
   Upon the different complexions and constitutions of men: That seems bitter to one, which is sweet to another (ibid.)
   Upon the difference of their external situation and position: Colours reflected from the clouds change according to the distance of the clouds, and according to the angle they make with the eye and luminous body. Fire also communicates the sensation of pleasure at one distance, and that of pain at another (ibid.)
(2) the same object cannot, at the same time, be endow'd with different qualities of the same sense (*Treatise* 1.4.4.4).
(3) the same quality cannot resemble impressions entirely different (ibid.)
(4) Instances of this kind are very numerous and frequent (*Treatise* 1.4.4.3)
(5) many of our impressions have no external model or archetype (*Treatise* 1.4.4.4)
   when different impressions of the same sense arise from any object, every one of these impressions has not a resembling quality existent in the object (ibid.)
   Many of the impressions of colour, sound, &c. are confest to be nothing but internal existences, and to arise from causes, which no ways resemble them (ibid.)
(6) These impressions are in appearance nothing different from the other impressions of colour, sound, &c. (ibid.)
(7) from like effects we presume like causes (ibid.)
(8) they are, all of them, deriv'd from a like origin (ibid.)
(9) colours, sounds, tastes, smells, heat and cold [. . . are] nothing but impressions in the mind, deriv'd from the operation of external objects, and without any resemblance to the qualities of the objects.
                                                              (*Treatise* 1.4.3.3)

The argument has a two-part structure. The first part consists of propositions (1) to (5) and the second of (6) to (9). In the first part, the goal is to argue that there are many impressions of proper sensibles that do not resemble their causes in their supposed Real objects. The second builds on this conclusion

Hume on the Distinction between Primary and Secondary Qualities 251

with an inductive-causal argument to the result that none of the impressions of proper sensibles resembles their supposed Real objects (PSP).

We can see the relevant points of the argument and Loeb's and Garrett's criticism by using an example from Locke (*Essay* II. viii. 21). It is closely similar to Hume's own illustration of fire producing the feeling of pain close at hand and the sensation of pleasure farther off. I just think that Locke's example is more telling. So, consider the case where you sink your hands into a bowl of warm water. One of your hands is cold and the other warm before sinking them. With the cold hand, the water feels warm, while in the warm hand the tactile impression is cold(er). In more general terms, here we have a case of two tactile impressions of sensation of one and the same object, or presumably more precisely, of a causal factor in it (the power of the water to produce feelings of temperature in the perceiver).

The first part of the argument (1–5) is arguing that only one of the perceptions of warmth and cold can resemble the causal factor in the water. This happens through inferring first that the water cannot be both warm and cold at the same time (proposition 2). The hidden premise for this conclusion is, as Wright observes, that warm and cold are contrary properties: x cannot be both warm and cold at the same time (although it does not need to be one of them) (Wright 1983: 109). It follows from this, as proposition (3) says, that the causal factor in the water cannot resemble both perceptions. In order to justify this further conclusion, it must be presupposed that the resemblance in question is specific, i.e., it is about a determinate qualitative temperature and not qualitative temperature in general.[26] If the cause in the water can be only either warm or cold, it is not possible that it is specifically similar to both as a determinate temperature.

In the conclusion of the first part of the argument, Hume does not actually do anything else but turns this around. As the causal factor in the water cannot specifically resemble both warmth and cold, it is also so in the other direction. At least one of these perceptions is not thus specifically similar to the causal factor. For the sake of simplicity, let us suppose that it is the feeling of warmth although Hume expresses it more generally in proposition (5). For that general statement, he needs proposition (4): there are many instances of this type. We can, however, focus on this particular case: the feeling of warmth does not specifically resemble the causal factor in the water.

Though it is not explicit in the text, the second part of Hume's argument consists of two phases. In the first causal phase, his intent is to extend the result of the first part to other actual perceptions. In our example, this means that the perception of cold is not similar to the causal factor in the water either. For reaching this conclusion, Hume uses the second part of his fourth rule of causal reasoning (proposition 7): from similar effects we infer similar causes (*Treatise* 1.3.15.6). How is this rule supposed to do the needed job?

Hume's argument seems to run as follows. At this point, we know that the feeling of warmth does not specifically resemble its cause in the water. Since warmth and cold are similar on a more general level as qualitative temperatures, we are supposed to use the fourth rule. From these premises, it seems to follow that their causes are similar on a more general level, too. Thus, they, or actually the causal factor in the water, are not of the *kind* of qualitative temperature. The feeling of cold cannot therefore resemble it either, neither as a determinate feeling of temperature nor as belonging to the category of qualitative temperatures.

The inductive use of the fourth rule is to generalize this conclusion to any perception of proper sensibles. First, we need to remember that perceiving proper sensibles varies a lot, of which this is only one illustration. Secondly, as all the perceptions of proper sensibles are of the same type, the inductive use of Hume's fourth rule is supposed to warrant extending the result of the causal phase of the second part to every possible perception of proper sensibles. They have similar causes in respect of not resembling the perceptions of proper sensibles.

For the purposes of this chapter, however, it is the causal phase of the second part that is relevant. The problem lies there. In the first place, Loeb objects that from the specific dissimilarity of the perception of warmth to the causal factor in the water, it does not follow that the feeling of cold does not resemble it either. Hume's mistake is to dismiss specific differences for the sake of more general similarities (Loeb 2002: 221). Loeb's conclusion is thus that the argument for the PSP "is far from just and regular"; Hume is hasty in his endorsement of it (ibid.: 222).

In order to understand Loeb's criticism, we must reconsider Hume's argument. We can focus on the properties of the water and the perceptions of it instead of their possible resemblance.

The result of the first part of the argument is that the causal factor in the water cannot be both warm and cold. We made the supposition that the water is not warm. The problem lies in inferring from this lack of a determinate property that the water cannot have *any* property of that determinable kind. From the fact that x does not have a particular property of the determinable kind K, it does not follow that it cannot have any property of the kind K. The water not being warm does not warrant us to conclude that it cannot have some other qualitative temperature, cold, for example. (Even in the case that we perceive it to have these different temperatures.) At least without some extra premises, it is entirely possible that in the sense-variation cases of proper sensibles, one of our perceptions resembles its cause in a Real body.

Hume's fourth rule on a more specific level does not help either. The cause of the feeling of warmth in the water is not itself warm. According to the fourth rule, the cause of the perception of cold must be similar to the cause of the perception of warmth (perceptions are similar as temperatures). Thus, it must be of the kind "not-warm" (in the more general argument, the conclusion is "not qualitative temperature"). The problem is that all other qualitative temperatures belong to that kind—cold included.

Hume on the Distinction between Primary and Secondary Qualities 253

Hence Loeb thinks that Hume's argument for the PSP is fundamentally flawed. He writes that it is as if Hume has forgotten his sixth rule of causal reasoning, according to which the difference in the effects must be due to the difference between their causes (*Treatise* 1.3.15.8; Loeb 2002: 221). However, it is not certain that rule six might have made Hume see the flaw of his argument. All we can infer by the sixth rule in the water case is that there is a difference in the total cause of the different sense-impressions. The causal factor in the water is an invariant circumstance in the total cause. The sixth rule alone does not help us to see the problem of the argument.

A more fundamental reply to Loeb's criticism can be detected by reflecting on another point he makes. Loeb wonders why Hume indicates that the argument was common among modern philosophers (EHU 12.1.15; *Treatise* 1.4.4.3). Neither Locke nor Berkeley advances it (Loeb 2002: 218–220). It is true that Hume's argument appears to be causal when it is considered on its own, whereas Berkeley's argument for the PSP is grounded on the notion of arbitrariness. Faced with the sense-variation regarding the perceptions of proper sensibles, it would be arbitrary, Berkeley claims, to prefer one to another. In order to avoid this ungrounded preference, we have to conclude that Real body does not have any proper sensible properties whatsoever—if there are Real bodies (Berkeley 1998: 73; Berkeley 1948–1957, 2: 186).[27]

Although this is how Loeb reads Berkeley, he does not see how Hume's argument can be read in the same way as Berkeley's argument (Loeb 2002: 220). However, recall that the second part of Hume's argument works on the parity of proper sensibles as qualitative properties. Could Hume's point be that their causes in Real bodies are also, by the fourth rule, on a par as dissimilar to their effects? Is it possible to read the argument in the way that preferring some proper sensible quality as similar to its Real cause would be arbitrary—given that the experience of the variation and the rules of causal reasoning are everything that we have?

Perhaps this is a possible reading of the argument and it calls for further research, for which there is not space here. Nevertheless, the argument does not explicitly speak about arbitrariness or anything concomitant to it. Therefore it seems to me that we ought to prefer Loeb's interpretation of the argument. On that reading, his objection to the argument hits the target. Hume's argument for the PSP really involves a gap.

After reconstructing the argument and observing a real hole in it, we can now discuss whether there is also another gap as Garrett suggests. If he is correct, it may give us a reason to doubt Hume's endorsement of the argument and the PSP—or at least the certainty of his belief in the PSP (Garrett 1997: 219). Garrett's point is that even if we succeeded in showing that the causal factor in the water is not itself of any qualitative temperature, it would not prove that the water taken in its entirety cannot have some qualitative temperature. The argument concerns rather the causal factor of the water to produce certain sense-impressions in us than the water in its entirety.

It seems to be clear that the possibility of this gap—or that of the first—does not surface in Hume's formulation of the argument. This may suggest at least three things. The first possibility is that there is a hidden premise that closes the gap and from Hume's point of view, the argument goes through. For example, according to the corpuscularian hypothesis, the Real water consists only of the corpuscles and it is their movement that is the (partial) cause of our feelings of warmth and cold. There is not thus anything else in the water than the corpuscles and their movement that could resemble these perceptions. The second explanation is that Hume sees the gap and in his view, the conclusion of the argument, the PSP, is possibly false. The argument is merely a good probable argument. The third possibility is that Hume does not realise the problem and accordingly his argument might be defective but he just does not get it. Although we may wonder how a great philosopher can be blind to such a basic mistake, this is how things seem to be in the case of the gap observed by Loeb.

Garrett's objection is based on his claim that the second possibility is the correct interpretation here. It is possible for Hume to doubt the PSP although he takes the argument for it as "satisfactory," "regular," and "just" (*Treatise* 1.4.4.3, 15, 1.4.7.4). Garrett thus thinks that the dubious nature of Hume's argument in Garrett's opinion raises doubts about Hume's commitment to its conclusion (Garrett 1997: 218–219).

This is, in my view, the first of two problems in this worry that I take to be sufficient for answering it. In the first place, even if Garrett is right in his doubts about the argument, it alone does not constitute evidence for Hume taking it as such or as a merely probable argument, without further support from the text. As far as I can see, there is no such support; Hume appears to be either blind to the possibility of the gap or he presupposes something in the argument that rules it out.[28] It would be rather strange for Hume to think that an argument is possibly defective and therefore merely probable without making it explicit.

The second problem in Garrett's worry is connected to the first that there is no textual evidence for it. On the contrary, there is strong textual evidence against it in the very context of the argument. As I said above, Garrett acknowledges that Hume takes the argument to be a just but only probable argument. Normally he is a careful reader of Hume, but here it is unfortunate that he misses one important passage. I have already observed that in *Treatise* 1.4.4.4 Hume clearly claims that the PSP is the most convincing.

From these problems it also follows that the second possible interpretation of the potential gap located by Garrett cannot be correct.[29] In the *Treatise*, Hume does not consider the argument for the PSP as possibly defective and therefore a merely probable argument. Garrett's third challenge cannot thus shake the interpretation that Hume endorses the PSP.

Three other doubts about Hume's subscription to the PSP and its rationality have also been overcome. We have thus good enough reasons to believe that he embraces this principle. However, we have not yet settled the problem of how certain his belief in the PSP is. I think it is possible to give a definitive answer for the *Treatise*, but the evidence in the later texts is not decisive. As was seen just above, Hume claims in the *Treatise* that the PSP is convincing to the highest conceivable degree. According to what Garrett calls Hume's "Conceivability Criterion of Possibility," conceivability implies possibility (1997: 24). So no inductive claim can be more convincing than the PSP. Of course, it cannot be Humean intuitive or demonstrative knowledge. The PSP is a contingent truth. Still, it is a conclusion of what Hume calls a "proof" in contrast to a "probability," which is Garrett's view of it and the principle. According to this distinction, proofs are "arguments, which are deriv'd from the relation of cause and effect, and which are entirely free from doubt and uncertainty" (*Treatise* 1.3.11.2). As the PSP is the most certain, it must be totally free from uncertainty (excluding the possibility of falsehood implied by contingency). In the *Treatise*, Hume's view is then that the PSP is based on a firm inductive-causal proof.

In the later texts, Hume seems to be more reserved although he speaks about a discovery, proof, establishing certainty, and what we are able to establish. In the light of these texts, it is completely possible that Hume changed his view about the certainty of the PSP and of the argument for it. In them, he may take them as merely "probable"—Garrett might be right about Hume's mature position. Yet I think the evidence strongly suggests that the PSP is at least highly probable in Hume's mature view. So even he does not consider it as based on a "proof" any longer, he continues to embrace the PSP and its rationality.

CONCLUSION

In this chapter, I have defended two main theses. First, Hume has an insight into the core of the distinction between primary and secondary qualities in what I call the Proper Sensibles Principle: proper sensible qualities are not Real properties of material substance and Real bodies. Hume is right when he claims that most new philosophers who included material substance into their system endorsed this principle (as understood by Hume).

My second main thesis is that Hume embraces the Proper Sensibles Principle because it is well grounded in inductive-causal, Humean reason. I have shown that there is sufficient textual evidence for taking Hume to assent to the rationality of the PSP in his writings from different periods. It is also most likely that that rationality is of the inductive-causal, Humean type.

When Hume's endorsement of the PSP and its rationality is established, the argument that the second profound argument against the senses and *Treatise* 1.4.4.6–15 are really his own reasonings can really take off. Its implication is that Hume's metaphysics of body must be nonmaterialist and his philosophy non-Realist in general.[30]

NOTES

1. EHU will be cited by a series of Arabic numerals: two in the case of the first *Enquiry* (section and paragraph).
2. *Treatise* will be cited by book, chapter, section, paragraph.
3. Metaphysical Realism is the doctrine that there are perception-independent, absolutely external, and continued entities. Following Michael J. Loux, in order to distinguish Metaphysical Realism and Metaphysically Real entities from the other uses of "realism" and "real" (and its cognates), I write the former with the capital "R" (Loux 2002: 250). For the sake of brevity, I will henceforth drop "Metaphysical." See Hakkarainen 2007 for a study of Hume's attitude to Realism.
4. Garrett 1997: 208 and 234; 2004: 83 and 90; 2006: 167 and 171.
5. Somewhat surprisingly, the New Humeans John P. Wright and Galen Strawson believe that Hume adheres to the distinction although they deny his endorsement of the conclusion of the second profound argument (Wright 1995/86: 232; and 1983: 109–111; Strawson 2002: 237–241).
6. Here I side with Peter Kail's excellent discussion of the topic (2007: Ch. 7.2.1).
7. See also *Essay* II. viii. 14, 15.
8. Hume speaks of "colours, sounds, and other secondary qualities" and "the secondary and sensible qualities."
9. Aristotle 1956 is cited by *De Anim*. Bekker numbers.
10. The third occurrence of this term in *Treatise* 1.4.4 is in §11 where Hume seems to treat it as the synonym of "secondary qualities."
11. He also uses the term in the broader sense of meaning any sensible property (e.g. EHU 4.2.16 and 7.1.8).
12. The PSP is also one of the Epicurean sides of the new philosophy. According to David Sedley, Epicurus' view is that atoms are completely void of proper sensible qualities. Even if Epicurus seems to think that macroscopic objects have those qualities, atoms do not have color, hardness, or temperature, sound, taste, and smell (Sedley 2005: 379–382). They "have only the ineliminable properties of all body: resistance, size, shape and weight." (ibid.: 379) So if the PSP is confined to the ultimate Real bodies in Epicurus' view, this tenet may be attributed to him.
13. As Anstey shows in 2000: Ch. 1.
14. Malebranche 1997 is cited by Book number. Ch. number. section number: page number.
15. For the details of this incident, see Hume 1986.
16. Instead, those passages where Hume states his view that sounds, smells, and tastes are nonspatial do not count as evidence for his subscription to the PSP (e.g. *Treatise* 1.3.14.25 and 1.4.5.10). In these passages, his intention is argue that these proper sensible qualities do not have any spatial existence (extension or location). Whereas colors and tactile qualities may have, in these passages he makes the distinction between spatial and nonspatial proper sensibles. As we

can see, this distinction is not identical with the divide between primary qualities and proper sensibles.
17. Besides, even in the *Treatise* Hume sometimes uses the same rhetorical devices when he is reporting his own views (*Treatise* 1.1.1.10, 1.2.1.2, 1.2.2.4, 1.2.5.8, 1.2.6.7, 1.3.9.11, 1.3.12.1, and 13.14.10).
18. There are mind-independent, continuous, and external entities.
19. The same reply applies to Bricke's, Wright's, and Baxter's contention that Hume's argument for the mind-dependency of perceptions presupposes Metaphysical Realism.
20. Kail makes the same point (2007: 70).
21. I realised the importance of this point because of Kail's comments.
22. That the PQP concerns comprehensible properties explains how Hume may think that the PSP is independent from the PQP. The PQP does not follow from the PSP because there may be incomprehensible properties in addition to proper sensible properties and primary qualities.
23. Equivalence is ruled out by Hume's contention that the PSP is more fundamental than the PQP.
24. Recall that above I have argued against Millican that the overall textual evidence shows that *Treatise* 1.4.4 and EHU 12.15–16 are not *modus tollens* arguments against the PSP alone.
25. Bricke and Wright have also very brief sketches of the argument (Bricke 1980: 18–19; Wright 1983: 109). But because of their brevity, I mainly concentrate on Loeb and Garrett. Bricke and Wright agree with me that Hume endorses the argument.
26. Here I make a distinction between qualitative and quantitative temperature to distinguish the feeling of it from measuring it.
27. Russell uses the same argument to the result that we immediately sense "sense-data," which "depend upon the relations between us and the object" (Russell 1912: 16 and 8–11).
28. Although Loeb takes Hume's argument to be defective, he does not draw, rightly in my opinion, the conclusion that Hume rejects its conclusion. On the contrary, Loeb thinks that Hume endorses it (Loeb 2002: 222).
29. Whether the first or the third is the correct interpretation is an interesting question and it has philosophical implications. It is not, however, relevant for the purposes of this chapter.
30. I would like to express my deepest gratitude to Kenneth Winkler, Don Garrett, Eric Schliesser, Peter Kail, and Peter Anstey for corrections, comments, and criticism. I am also grateful for the invitation to and the comments of the audience at the workshop *Early Modern Epicureanism and Anti-Epicureanism*, University of Leiden, April 11, 2008.

BIBLIOGRAPHY

Algra, K., Barnes, J., Mansfeld, J., and Schofield, M., eds. (2005) *The Cambridge History of Hellenistic Philosophy*, Cambridge: Cambridge University Press.
Anstey, P. R. (2000) *The Philosophy of Robert Boyle*, London: Routledge.
Aristotle (1956) *De Anima*, ed. W. D. Ross, Oxford: Clarendon Press.
Baxter, D. L. M. (2008) *Hume's Difficulty—Time and Identity in the* Treatise, New York: Routledge.
Baier, A. (1991) *A Progress of Sentiments—Reflections on Hume's* Treatise, Cambridge MA: Harvard University Press.

Berkeley, G. (1948–1957) *The Works of George Berkeley*, 9 vols, eds. A. A. Luce and T. E. Jessop, London: Nelson
———. (1998) *Three Dialogues between Hylas and Philonous*, ed. J. Dancy, Oxford: Oxford University Press; 1st edn 1713.
Bricke, J. (1980) *Hume's Philosophy of Mind*, Edinburgh: Edinburgh University Press.
Cottingham, J. (1993) *A Descartes Dictionary*, Oxford: Blackwell.
Garber, D. (1992) *Descartes' Metaphysical Physics*, Chicago: University of Chicago Press.
Garrett, D. (1997) *Cognition and Commitment in Hume's Philosophy*, Oxford: Oxford University Press.
———. (2004) "'A Small Tincture of Pyrrhonism': skepticism and naturalism in Hume's science of man" in ed. W. Sinnott-Armstrong 2004, pp. 68–98.
Garrett, D. (2006) "Hume's Conclusions in 'Conclusion of This Book'" in ed. S. Traiger 2006, pp. 151–175.
Hakkarainen, J. (2007) *Hume's Scepticism and Realism—His Two 'Profound' Arguments against the Senses in* An Enquiry concerning Human Understanding, Tampere: Acta Universitatis Tamperensis.
Hume, D. (1986) "David Hume on Thomas Reid's 'An Inquiry into the Human Mind, on the Principles of Common Sense': a new letter to Hugh Blair from July 1762," ed. P. Wood, *Mind*, 95: 411–416.
———. (1987) *Essays, Moral, Political, and Literary*, ed. E. F. Miller, Indianapolis: Liberty Fund; various editions 1741–1777.
———. (2000) *An Enquiry concerning Human Understanding*, ed. T. L. Beauchamp, Oxford: Clarendon Press; 1st edn 1748.
———. (2007) *A Treatise of Human Nature*, eds. D. F. Norton and M. J. Norton, Oxford: Clarendon Press; 1st edn 1739–1740.
Kail, P. (2007) *Projection and Realism in Hume's Philosophy*, Oxford: Oxford University Press.
Livingston, D. W. (1984) *Hume's Philosophy of Common Life*, Chicago: University of Chicago Press.
Locke, J. (1975) *An Essay concerning Human Understanding*, 4th edn, ed. P. H. Nidditch, Oxford: Clarendon Press; 1st edn 1690.
Loeb, L. (2002) *Stability and Justification in Hume's* Treatise, Oxford: Oxford University Press.
Loux, M. J. (2002) *Metaphysics—a Contemporary Introduction*, 2nd edn, London: Routledge; 1st edn 1998.
Malebranche, N. (1997) *The Search after Truth*, trans. and eds. T. M. Lennon and P. J. Olscamp, Cambridge: Cambridge University Press; 1st edn 1674–1675.
Miller, E. F. (1987) "Foreword," in Hume 1987, pp. xi–xviii.
Millican, P., ed. (2002a) *Reading Hume on Human Understanding*, Oxford: Oxford University Press.
———. (2002b) "The context, aims, and structure of Hume's First Enquiry" in ed. P. Millican 2002a, pp. 27–65.
Morris, W. E. (2000) "Hume's Conclusion," *Philosophical Studies*, 99: 89–110.
Reid, T. (1764) *An Inquiry into the Human Mind on the Principles of Common Sense*, Edinburgh.
Richman, K. A. and Read, R., eds. (2007) *The New Hume Debate*, London: Routledge; 1st edn 2000.
Russell, B. (1912) *The Problems of Philosophy*, Oxford: Oxford University Press.
Sedley, D. (2005) "Hellenistic physics and metaphysics" in eds. K. Algra et al. 2005, pp. 355–411.

Sinnott-Armstrong, W., ed. (2004) *Pyrrhonian Skepticism*, Oxford: Oxford University Press.

Strawson, G. (2002) "David Hume: objects and power" in ed. P. Millican 2002a, pp. 231–257.

Traiger, S., ed. (2006) *The Blackwell Guide to Hume's Treatise*, Oxford: Blackwell.

Wright, J. P. (1983) *The Sceptical Realism of David Hume*, Minneapolis: University of Minneapolis Press.

———. (1986) "Hume's academic scepticism: a reappraisal of his philosophy of human understanding," *Canadian Journal of Philosophy*, 16: 407–435.

Contributors

Vlad Alexandrescu is Professor of Early Modern Studies at the University of Bucharest.

Peter R. Anstey is Professor of Early Modern Philosophy at the University of Otago, New Zealand.

Roger Ariew is Professor of Philosophy at the University of South Florida.

Katherine Brading is William J. and Dorothy K. O'Neill Collegiate Professor of Philosophy at the University of Notre Dame.

Sorin Costreie is Lecturer in Philosophy at the University of Bucharest.

Mihnea Dobre is Post-Doctoral Researcher at the Center for the Logic, History and Philosophy of Science, University of Bucharest.

Daniel Garber is Stuart Professor and Chair of Philosophy at Princeton University.

Jani Hakkarainen is Post-Doctoral Researcher, Academy of Finland, at the Department of History and Philosophy, University of Tampere, Finland.

William Harper is Professor of Philosophy at the University of Western Ontario.

Dana Jalobeanu is Lecturer in Philosophy at the University of Bucharest.

Gemma Murray is an MPhil candidate at St John's College, University of Cambridge.

Eric Schliesser is BOF Research Professor, Philosophy and Moral Sciences, Ghent University.

Curtis Wilson is Emeritus Professor at St. John's College in Annapolis, Maryland.

Index

A

Ablondi, F. 28 n. 15, n. 20, n. 21
absolute certainty 31, 32, 37–41, 43
Académie des Sciences 161, 169
accelerative force 93, 228
action-at-a-distance 83, 85, 86, 88, 90–2, 97, 98, 150, n. 18, 215, 225, 226, 229
Adams, R. 212 n. 7
aggregates: of atoms 6, 51, 53, 54; and mass 205; of monads 198, 199–203, 206, 208, 209, 212 n. 12; of parts 144, 221, 232 n. 7; as substances 196, 218, 219; *see also* Leibniz
air resistance 182–4, 187
Alexandrescu, V. 4
anatomy 4, 62
Anstey, P. R. 4, 76 n. 1, 96 n. 20, 121 n. 5, 256 n. 13, 257 n. 30
Arianism 81
Ariew, R. 3, 31, 50
Aristotelianism 130, 196, 227
Aristotle 4, 6, 35, 46, 71, 73, 75, 115, 126 n. 43, 231 n. 5, 256 n. 9; four element theory 63, 64, 69; on the infinite 232 n. 5; on place 56; prime matter 75; substantial forms 227; theory of qualities 4, 52, 63–5; 238; against void 215
Aristotelians 27 n. 4, 196
Arnauld, Antoine 55, 59 n. 10; and Leibniz 195–8, 205, 206, 208, 211 n. 3, 212 n. 9
Arriaga, Roderigo 32, 40–3
Arthur, R. 232 n. 5
astronomy 32, 34, 75, 94, 114; phases of Venus 32
atheism 81

atomism 3, 4, 14, 15, 18, 19, 22, 26, 36, 51, 53, 57, 65, 66, 73, 74, 75, 77 n. 8, n. 9, 104, 105, 140, 144, 215; *see also* Bernier, Cordemoy, Democritus, Descartes, du Roure, Epicureanism, Epicurus
atoms 3, 14, 16, 19, 22, 26, 50–5, 58 n. 7, 65, 70, 71, 72, 74, 97 n. 28, 120, 124 n. 29, 141, 195, 196, 198, 215, 216, 218, 221, 222, 223, 228, 230, 256 n. 12; *see also* atomism, monads
attraction 82, 84–8, 92–4, 95 n. 6, 225, 228; *see also* gravity

B

Bacon, Francis 62, 69, 77 n. 17, 103, 120 n. 1; on crucial experiments 228
Baconianism 103, 104, 105, 106, 120 n. 1
Baier, A. 236
Baillet, Adrien 123 n. 27
Basso, Sebastien 27 n. 4
Baumgarten, Alexander 213 n. 24
Baxter, D. 247, 257 n. 19
Bayle, Pierre 2, 7 n. 1, 213 n. 24
beauty 242–4, 246
Beeckman, Isaac 27 n. 4
Belkind, O. 87, 93, 98 n. 37, n. 39
Bentley, Richard 81, 83, 87, 89, 90, 91, 92, 96 n. 22
Berkeley, George 1, 2, 213 n. 24, 235, 237, 253
Bernier, François 4, 49–59 *passim*; *Abrégé* 49–57; attack on Descartes 50–5; on atomism 49–5; on motion 56, 59 n. 12; on space 55–7
Bernoulli, Johann 212 n. 11, 216, 219, 232 n. 11

264 *Index*

Bertoloni Meli, D. 97 n. 30, 122 n. 13
Biener, Z. 96 n. 20, 98 n. 39
Birch, Thomas 112, 124 n. 30, n. 31, 189 n. 15
Blackwell, R. 188 n. 4, 189 n. 8, n. 10, n. 12
Blair, Hugh 241–2, 244, 246
Blake, R. M. 31
body 1, 5, 11–26 *passim*, 53–56, 116, 117, 119, 126 n. 41, 136, 138, 144–6; as aggregates of monads 199–201; Descartes on 11–18, 108–9, 123 n. 22, 132, 141, 235; Hooke on 126 n. 41; Hume on 235, 236, 240, 256; Leibniz on 195–211 *passim*, 212 n. 18, 223; lone bodies 4, 80–94, 97 n. 28, n. 32, 119, 133–4, 148; Newton on 89, 132; simple bodies 80, 97 n. 28; and space 95 n. 6; *see also* individuation, matter
Borelli, Giovani Alfonso 61, 112, 124 n. 30
Borelli, Petrus 154
Boscovich, Roger J. 2
Bourguet, Louis 203
Boyle, Robert 1, 4, 61, 66–8, 69, 70, 76, 77 n. 10, n. 16, 114, 119–20, 124 n. 30; Boyle's Law 1; *Forms and Qualities* 1, 67–8, 105; 'An Invitation' 66; *Sceptical Chymist* 67; *Usefulness of Natural Philosophy* 64, 67, 68, 74, 111
Brading, K. 5, 97 n. 28, 98 n. 38, n. 39, 132, 134, 138, 149 n. 1,
Brahe, Tycho 32
Bricke, J. 247, 257 n. 19, n. 25
Brouillette, S. 97 n. 22, 98 n. 39
Brouncker, William 114, 124 n. 31, 155
Brown, G. 227, 228

C

Cantor, G. 224, 232 n. 5
Carlin, L. 232 n. 5
Cartesians 3, 18–22, 25, 26, 29 n. 30, 31, 43, 44, 49, 55, 57, 103–10, 120, 126 n. 41, 140, 196, 199, 220
Castle, George 4, 69, 70–2, 76, 77 n. 14, n. 15
Charles II 114
Charleton, Walter 4, 57 n. 2, 66, 77 n. 7, n. 9
Chauvin, Etienne 45 n. 8

Christina, Queen of Sweden 39
chymical philosophy 72, 73
chymical physicians 62, 65, 66
circulation of the blood 62
Clagett, M. 189 n. 17
Clarke, D. 29 n. 33, 31
Clarke, Samuel 6, 81, 83, 86, 92, 95 n. 6, 226
Clericuzio, A. 76 n. 2
Clerselier, Claude 40, 109, 123 n. 18, n. 25
clock metaphor 25–6, 36, 72
coefficient of restitution 172, 187
Cohen, I. B. 96 n. 9, 151 n. 22
cold, *see* qualities, primary qualities (Aristotelian *primae qualitates*), primary and secondary quality distinction
College of Physicians 70
Collins, John 112
collisions: laws of 5, 6, 55, 103–20 *passim*, 133–9, 146, 147, 149 n. 9, n. 12, 150 n. 14, n. 15, 153–90 *passim*, 227, 231; elastic collisions 159–61, 171–3, 189 n. 8; imperfectly elastic collisions 182–87; inelastic collisions 175–81
composite systems 130–48 *passim*
Conimbricences (Jesuits of Coimbra) 42
conservation of kinetic energy 106, 159, 160, 188 n. 3, 190 n. 20
conservation laws 133–7, 148, 149 n. 6, n. 7, n. 9, n 12; 150 n. 12; *see also* Descartes, Newton
conservation of momentum 147–8, 150, n. 14, 159–61 172, 180, 187 n. 3, 188 n. 3; *see also* impetus
conservation of motion 113, 116, 125 n. 38, n. 39, 134–6, 146–8, 149 n. 6, n. 12, 150 n. 14
Conti, Antonio-Schinelli 228
contiguity 53, 54, 58 n. 6, 197
continuum 6, 51–4, 57 n. 2, 124 n. 29, 198, 202, 215, 216, 218–25, 232 n. 6, n. 7
contraries, doctrine of 64–5, 69
Cook, H. 76 n. 2, n. 3, n. 4
Copernicus, Nicolaus 32
Cordemoy, Gerauld de 3, 11, 12, 28 n. 10, n. 15, n. 16, n. 17, n. 18, n. 20, n. 22, n. 24, n. 29, n. 30, 196; on atomism 18–21, 22, 26, 196

corpuscular matter theory 4, 61, 65, 66, 68, 70, 71, 72, 73, 74, 235, 241, 254
corpuscularianism 67, 68, 71, 72, 73, 76, 105, 235, 241
Costreie, S. 6
Cotes, Roger 91, 92, 97 n. 27
Cottingham, J. 240
Cover, J. A. 232 n. 15
creation 33, 87, 108; of motion 113
Croone, William 124 n. 30
Curley, E. 45 n. 8

D

Dagnicourt, Pierre 205
de Volder, Burchard 199–211, 212 n. 13, n. 14, n. 19
Debus, A. 76 n. 2
definition of body 5, 14, 15, 16, 25, 28 n. 21, 108, 111, 116, 117, 119, 120, 126 n. 41, 132, 205, 223
definition of motion 3, 11, 16, 19, 22, 25, 28 n. 21, 56, 57, 59 n. 12, 108, 111, 120, 131, 132
Democritus 22, 36, 199, 204
density 13, 14, 27 n. 4, 108, 110, 221, 224
Desargues, Gérard 51
Des Bosses, Bartholomew 199–202, 206–11, 212 n. 15, n. 17, n. 18
Descartes, René 1–5, 11–29 *passim*, 51–5, 65, 68, 75, 81, 82, 87, 104, 105, 107–9, 117, 122 n. 11, n. 15, 123 n. 29, 125 n. 39, 130, 138, 139, 147, 149 n. 1, n. 2, 154, 196, 235, 240; animals 38–9, 61; against atomism 14, 15, 16, 27 n. 8, 140–1; on collision laws 5, 107–9, 122 n. 18, 135, 149 n. 12, 170, 189 n. 12; on condensation 14; conservation laws 133–7, 149 n. 6, n. 7, n. 10, n. 11, n. 12, 150 n. 12, n. 14; *Dioptrics* 33, 34; *Discourse* 3, 33, 34, 38–9, 40, 44 n. 2, 122 n. 2; on divisibility of matter 3, 15, 16, 19, 28 n. 10, 51, 53, 55, 132–3, 140–3, 145; eternal truths 39; fluids 21, 22, 23 29 n. 25; on hypotheses 3, 31–44; illustrations, Descartes's use of 12, 15–17, 20, 22–25, 28 n. 14, 29 n. 32; on indefinite space 15, 19; on individuation 2, 3, 11–26 *passim* 132, 134–6, 137, 141, 148, 149 n. 2, n. 3; laws of motion 107–9, 132; laws of nature 54, 131–6, 142, 149 n. 9; on matter as extension 3, 11, 13–15, 19, 20, 27 n. 4, 87, 107, 131, 196; *Meteors* 22–3, 33, 34; *Meditations* 12, 35, 39, 40, 44 n. 5, 53, 54; on moral certainty 2, 3, 31, 32, 36–41, 43; on motion 3, 11, 15–22, 25, 27 n. 10, 28 n. 14, n. 21, 32, 33, 59 n. 12, 107–9, 123 n. 22, n. 23, n. 25, 125 n. 37, 131–4; on physiology 61, 107; on planetary motion 21–2; *Principles* 1, 3, 11, 12, 13, 16, 21, 24, 25, 27 n. 1, 31–8, 51–3, 58 n. 4, 107, 108, 122 n. 16, 123 n. 20, 131, 133, 135–6, 140–3, 149 n. 6, n. 8, 150, n. 12, 189 n. 12; on rarefaction 13–14, 27 n. 4; *Rules* 44 n. 3; on sensible qualities 12–13; *Specimina* 34, 40, 44 n. 4; *Treatise on Man* 61; on the void 13, 14, 27 n. 7, 36; *The World* 34–5, 133–5, 149 n. 6, n. 7, n. 9, 150, n. 12
Des Chene, D. 27 n. 4, 122 n. 17, 123 n. 23, n. 26
Des Maizeaux, Pierre 213 n. 20
determination of motion; *see* motion (direction)
Diderot, Denis 213 n. 24
disease: Galenic theory 63–6; miasmic theory 70; seminal theories 70, 77 n. 13; wormatic theory 70, 71
dispositional qualities 80–93 *passim*
divisibility 3, 6, 15, 16, 19, 22, 28 n. 10, n. 19, 29 n. 27, 36, 51, 53, 55, 58 n. 3, 94, 133, 140–6, 197–8, 215–21 *passim*, 230
Dobre, M. 3
Downing, L. 96 n. 21
dryness 50, 52, 63, 69, 71, 72; *see also* primary qualities (Aristotelian *primae qualitates*)
Du Roure, Jacques 3, 12, 22, 26, 29 n. 26, n. 27
Ducasse, C. J. 31

E

Eisler, R. 213 n. 24

elasticity 109, 110, 111, 113–16, 126 n. 41, 139, 181, 187, 228; *see also* collisions (elastic collisions), (imperfectly elastic collisions)
Empiricism 82
Empiricists 81, 82
Epicureanism 91, 92, 227, 241
Epicurus 22, 75, 90, 91, 196, 213 n. 24, 256 n. 12
essential attributes and properties 11, 53, 80, 87–92, 96 n. 22, 97 n. 22, n. 27, n. 32
ether theory of gravity 82, 86
Eustachius a Sancto Paulo 32, 41–3
experiment 2, 6, 11, 39, 62, 68, 69, 76, 82, 84, 94, 98 n. 38, 103, 105–107, 110–12, 116, 117, 120, 121 n. 2, n. 3, n. 4, 124 n. 30, n. 31, n. 32, 125 n. 34, 126 n. 41, 153, 154, 171, 172, 175, 182–7, 190 n. 22; crucial experiments 228; thought experiments 14, 27 n. 5, 161–9; *see also* pendulum experiments
Experimental Philosophy 77 n. 17, 84, 103, 121 n. 2, n. 4, n. 5, 199, 228
extension 2, 3, 11–15, 19, 20, 27 n. 4, 56, 87, 95 n. 6, 107, 126 n. 41, 131, 132, 195, 196, 200, 202, 203, 206, 207, 212 n. 19, 218, 221, 239, 240, 256 n. 16

F

Fabri, Honoré 154
faith 41–2, 207
Familiarity Condition 3
Fatio de Duillier, Nicolas 90, 97 n. 31
fermentation 90
Fichant, M. 122 n. 14, n. 15
field 80, 83, 85, 92, 93, 96 n. 15
fluidity 20–1, 29 n. 25, 52, 68, 71, 120, 126 n. 41, 218
force 5, 80, 83, 85–95, 95 n. 5, 96 n. 11, 97 n. 30, 106, 108, 109, 111, 113, 119, 120, 122 n. 17, 123 n. 18, 126 n. 39, 134, 137, 138, 144, 148, 150 n. 12, n. 14, 151 n. 22, 172–86, 188 n. 3, 189 n. 12, n. 18, 190 n. 20, 199–208, 220, 226, 228, 230, 231, 231 n. 3
form 196, 197, 206, 207, 209, 212 n. 11, 213 n. 24, 230; *see also* substantial forms

Foucher, Simon 218
Frank, R. G. Jr 76 n. 3, 77 n. 8, n. 10
Freddoso, A. J. 122 n. 17
Friedman, M. 98 n. 39

G

Gabbey, A. 122 n. 18, 149 n. 12
Gale, G. 230
Galen 4, 65, 71, 73
Galenic medicine 4, 62–76 *passim*; humoral theory 4, 63, 65, 71–3, 75, 77 n. 14; temperaments 63, 70, 71, 73, 77 n. 14
Galileo 1, 35, 82, 123 n. 29, 154, 170
Galison, P. 29 n. 31
Garber, D. 6, 11, 25, 27 n. 3, 28, n. 11, n. 19, n. 21, n. 23, 31, 44 n. 2, n. 3, 122 n. 16, 123 n. 23, 149 n. 6, n. 8, n. 9, n. 12, 150 n. 12, n. 17, 211 n. 1, n. 2, 212 n. 6, n. 8, 220, 227, 231 n. 2, n. 3, 240
Garrett, D. 236, 237, 247–55, 256 n. 4, 257 n. 25
Gassendi, Pierre 3, 4, 22, 49–57 *passim*, 57 n. 1, 59 n. 9, n. 12, 65, 66, 74, 75, 77 n. 9, n. 16, 104–6, 120, 122 n. 10, 123 n. 29; on space 56; *Syntagma* 49–57 *passim*
Gaukroger, S. W. 11, 27 n. 1, 44 n. 2, 123 n. 21, n. 23, 149 n. 7, 150 n. 12
generation 105, 221
Gerhardt, C. I. 212 n. 3
Glanvill, Joseph 104–5, 110, 122 n. 11, 124 n. 29
God 14, 18, 25, 27 n. 6, n. 7, 33, 36–8, 40–3, 44 n. 2, 56, 80, 82, 87–90, 95 n. 5, 98 n. 35, 108, 109, 125 n. 38, 140, 149 n. 2, 198, 207, 212 n. 18, 216, 223, 225–7, 229, 232 n. 5; God's absolute power (*potentia dei absoluta*) 27 n. 6, n. 7, 36, 40, 41, 43, 45 n. 8; God's attributes 82; omnipresence 98 n. 35; God's ordinary power (*potentia ordinata*) 51
gravity 4, 6, 80–95 *passim*, 125 n. 36, 225–9; as cause 85; cause of 80, 83, 85–7, 89–91, 93, 96 n. 22; as (non)essential 88–9, 92, 97 n. 32; force of 80, 83, 85, 93; law of 89; as relational

quality 83–90; *see also* attraction, 'ether theory of gravity, Leibniz, medium
Grene, M. 18
Grosholz, E. 11
Gueroult, M. 150, n. 12

H

Habert de Montmor, Henri Louis 49
Hakkarainen, J. 6, 235, 256 n. 3
Hall, A. R. 122 n. 13, 124 n. 31, 172, 189 n. 6, n. 11, n. 14, n. 15, n. 16, n. 18
Hall, M. (Boas) 121 n. 1, 122 n. 13, 124 n. 31
hardness 29 n. 25, 51–4, 88, 109–11, 113–16, 126 n. 41, n. 45, 139, 150, n. 15, 156, 157, 160, 161, 170, 179, 181, 182, 186, 189 n. 8, n. 12 237, 239, 256 n. 12; *see also* solidity
Harper, V. 187 n. 1
Harper, W. 5, 98 n. 39, 139, 187 n. 1
Hartley, H. 114
Hartz, G. A. 232 n. 15
Harvey, William 62
Hattab, H. 122 n. 17
Hearne, Thomas 115
heat 52, 63, 64, 69, 71, 72, 238, 241, 244, 250; *see also* primary qualities (Aristotelian *primae qualitates*)
Helmont, Joan Baptiste van 62, 64, 69, 77 n. 13
Henry, J. 87, 89, 91, 96 n. 21, 97 n. 28, 98 n. 39
Herivel, J. 150, n. 13, n. 20
Highmore, Nathaniel 66, 77 n. 8
Hippocrates 65, 69, 72, 73
Hobbes, Thomas 1, 92, 123 n. 29, 196
Holden, T. 139–43, 149 n. 2
Hooke, Robert 81, 110, 111, 114–16, 124 n. 30, n. 31, 126 n. 41
Hume, David 1, 2, 6, 82, 95 n. 5, 235–56 *passim*; *Enquiry* 1, 95 n. 5, 235–47 *passim*, 249, 253, 256 n. 1, n. 11, 257 n. 24; on impressions 238, 245, 250–1, 253, 254; on primary and secondary quality distinction 235–56, 256 n. 8, n. 10; *The Sceptic* 243, 244, 246, 249; skepticism about matter 6, 82, 235, 256; *Standard of Taste* 242–3, 246, 249; *Treatise* 235–56 *passim*, 256 n. 2, n. 10, n. 16, 257 n. 17, n. 24
Hunter, M. 67, 77 n. 10, n. 16, 120 n. 1, 121 n. 4, 126 n. 44
Hutcheson, Francis 244–5, 246
Huygens, Christiaan 3, 5, 6, 81, 82, 106, 107, 109–15, 119, 123 n. 27, n. 28, 124 n. 31, n. 33, 125 n. 34, n. 35, 150 n. 12, 153, 172, 187 n. 3, 188 n. 4, n. 5, 189 n. 6, n. 8, n. 10, n. 12, 228; collision laws 154–70, 173, 182, 186, 187; hypotheses 44, 189 n. 7, n. 9; rules of motion 156–70; *Treatise on Light* 43–4; on weight 157, 160, 188 n. 4
Huygens, Constantijn 39
hylomorphic theory of matter 4
hypotheses 2, 3, 31–44 *passim*, 85, 94, 103–20 *passim*, 120 n. 1, 121 n. 4, n. 5, 122 n. 12, 123 n. 27, 124, n. 31, n. 32, 125 n. 37, n. 39, 126 n. 41, 144, 155, 170, 227–8; *see also* Huygens (Christiaan), Newton, suppositions
hypothetico-deductive method 31, 43

I

iatrochemistry 4
idealism 211, 213 n. 24, n. 25, 247, 256
ideas, clear and distinct 18, 20, 24, 36, 54, 131, 149 n. 2, n. 3
illustrations, *see* Descartes
imagination 15, 18–22, 26, 121 n. 5
impenetrability 19, 28 n. 18, 52, 88, 205, 206, 228
imperfectly elastic collisions, *see* collisions
impetus 126 n. 45, 176–81, 228
impressions, *see* Hume
Index of Prohibited Books 3, 50
individuation of bodies, *see* Descartes, Newton
indivisibility, *see* divisibility
induction 120 n. 1, 121 n. 1, 228, 229, 236, 245, 252, 255
inertia 97 n. 30, 106, 108, 113, 116–19, 161
infinite 19, 27 n. 9, 28 n. 10; actual 219, 223, 224, 232 n. 5, n. 6; potential 219, 232 n. 5; syncategorematic 217, 218, 231 n. 5; *see also* divisibility

268 Index

innate ideas 44 n. 2, 54, 229
instrumentalism 75–6, 81, 83, 95 n. 6

J

Jalobeanu, D. 5, 96 n. 20, 98 n. 39, 103, 105, 121 n. 6, 122 n. 17, 123 n. 19, n. 26, 139, 149 n. 4, n. 5, 150 n. 19
Janiak, A. 80, 82–6, 88–92, 94, 95 n. 6, 96 n. 11, n. 16, n. 18, n. 19, n. 22, 97 n. 27, n. 32, 98 n. 34, n. 35, 99 n. 39, 150 n. 18
Jaquelot, Isaac 213 n. 20
Jones, R. F. 77 n. 11
Jonson, Ben 115, 126 n. 43
Joy, L. 97 n. 29, 122 n. 10
Jungius, Joachim 125 n. 34, 126 n. 40, 154

K

Kail, P. 256 n. 6, 257 n. 20, n. 21, n. 30
Kant, Immanuel 81, 97 n. 27
Kaplan, B. B. 77 n. 10
King, L. 77 n. 11
Kochiras, H. 98 n. 34
Koyré, A. 49

L

Lamy, Bernard 25
Laudan, L. 31
Law-constitutive approach 5, 130–48 passim
laws of collisions, see collisions, Descartes, Huygens, Wallis, Wren
laws of mechanics 35
laws of motion 1, 2, 5, 6, 90, 103–20 passim, 124 n. 29, n. 31, n. 34, 125 n. 35, n. 39, 130–48 passim, 153–90 passim, 228; see also collision laws, laws of mechanics, laws of nature
laws of nature 2, 5, 51, 54, 87–8, 95 n. 5, 107, 108, 131–6, 210, 226, 227, 229
Le Febvre, Nicaise 64
L'Hospital, Guillaume de 197
Leibniz, Gottfried W. 6, 11, 28 n. 10, 81, 83, 84, 86, 92, 94, 95 n. 6, 96 n. 18, 98 n. 35, 106, 195–232 passim; and Arnauld 195–8, 205, 206, 208, 211 n. 3, 212 n. 9; bodies as aggregates of monads 6, 199–201, 209, 212 n. 12; common dream view of bodies 6, 201–3, 209, 212 n. 18; bodies as composite substances 199, 205–6; identity of indiscernibles 225; against Newton's gravity 215, 225–9; and idealism 195, 211, 213 n. 24; Monadology 195, 199, 200, 213 n. 21; New System 197–8; pre-established harmony 213 n. 24, 229; principle of plenitude 216, 222; Principles of Nature and Grace 199, 200, 206, 213 n. 21; principle of sufficient reason 224, 225; on substance 6, 196–211 passim; on substantial bonds 6, 200, 202, 206–9, 210, 213 n. 22; on substantial forms 196–7, 207, 208, 227; Theodicy 199, 222; on the void 215–31 passim
Leibniz/Clarke correspondence 6, 81, 83, 86, 92, 95 n. 6, 226
Lennon, T. M. 122 n. 9, 142
lever 189 n. 18
Levey, S. 217, 221, 223, 224
Livingston, D. W. 236
Locke, John 1, 2, 6, 65, 66, 70, 77 n. 13, 82, 96 n. 17, 237, 238, 240, 241, 242, 244, 251, 253
Lodge, P. 212 n. 13, n. 19
Loeb, L. 237, 249–53, 254, 257 n. 25, n. 28
Look, B. 207, 212 n. 15, 213 n. 22
Loux, M. J. 256 n. 3
Lucretius 56
Lüthy, C. 29 n. 32

M

McClaughlin, T. 31
McGuire, J. E. 27 n. 9, 89, 96 n. 9, n. 13, n. 14, 98 n. 38, n. 39, 143
MacLaurin, Colin 84, 98 n. 33
McMullin, E. 96 n. 19
Madden, E. H. 31
Maignan, Emmanuel 56, 57, 59 n. 11
Malebranche, Nicolas 212 n. 4, 240, 241, 242, 244, 256 n. 14
Mandelbaum, M. 29 n. 29
Marci, Marcus 112
Mariotte, Edme 182, 189 n. 19
Masham, Damaris 212 n. 14, 225
mass 83–90 passim, 106, 108, 147, 156, 157, 158, 160, 173, 180, 188 n. 3, n. 4, 197, 200, 201, 204, 205

Mates, B. 231 n. 4, 232 n. 15
mathematics 6, 37, 39, 41, 84, 96 n. 11, 108, 109, 110, 113, 125 n. 39, 210, 216, 217, 219–21, 224, 230
matter theory 1–6, 116, 117, 130, 136, 138–9, 145–6, 150 n. 16; actual/potential parts debate 139–45; Aristotelian 6, 34; of Bernier 50–7; of Cordemoy 18–21; of Descartes 12–18, 25–6, 27 n. 1, 28 n. 11, 29, 33, 75, 131–2; of Gassendi 50–7; of Newton 96 n. 20, 130, 132, 136–8; *see also* divisibility, gravity, impenetrability, mass, qualities
Maxwell, J. C. 91, 98 n. 33
Mayow, John 61
mechanical affections 4, 6; *see also* primary qualities
mechanical philosophy 3, 4, 54, 61, 72, 74, 76, 83, 84, 96 n. 8, 107, 111, 120, 123 n. 29, 227
mechanics 2, 6, 11, 24, 27 n. 1, 35, 36, 104, 106, 107, 109, 123 n. 28, 124 n. 30, 137, 138, 145–6, 227; Newtonian 130, 138, 145–6; *see also* laws of mechanics, principles of mechanics
medicine 61–76 *passim*
medium theory of gravity 80, 85, 86, 88, 89, 91, 92, 97 n. 32; *see also* gravity
Menn, S. 122 n. 17
Merchant, C. 122 n. 14
Mersenne, Marin 27 n. 6, 34, 35, 39, 45 n. 6
Mesland, Denis 33, 39, 45 n. 6
metaphysical certainty 38–40, 42
metaphysics 6, 12, 35, 43, 44 n. 3, 81, 82, 83, 87, 113, 125 n. 38, 130, 131, 134, 137–46 *passim*, 149 n. 1, 197, 198, 209, 210, 213 n. 24, 224, 227, 229–30, 235–6, 240, 256
Miller, D. M. 98 n. 38, n. 39
Miller, E. F. 243
Millican, P. 246, 257 n. 24
Mind, *see* soul
miracles 51, 98 n. 35, 225, 226–7, 228
modes 14, 55, 68, 109, 123 n. 24, 125 n. 37, 200, 207

moistness 50, 52, 69, 71; *see also* primary qualities (Aristotelian *primae qualitates*)
momentum 86, 147, 150 n. 14, 159–61, 172, 180, 187 n. 3
monads 6, 195–213 *passim*, 227, 229, 230; *see also* Leibniz
Montmor, Habert de 49
Moore, A. 232 n. 5
moral certainty 2, 3, 31–45 *passim*
More, Henry 11, 26, 123 n. 24
Morin, Jean-Baptiste 34, 35, 38
Morris, W. E. 236
motion, definition of 3, 11, 15, 16, 19, 22, 25, 28 n. 21, 56–7, 59 n. 12, 108, 111, 120, 123 n. 22, n. 25, 132; destruction of 113; direction (determination) of 108, 118, 137, 147, 158, 159, 160, 168, 171, 172, 177, 179, 185; transfer of 15, 108, 113, 123 n. 24, 125 n. 37, 131, 172, 203; *see also* conservation of motion, laws of motion, principles of motion
Mouy, P. 44
Murr, S. 50
Murray, G. 5, 139

N

natural history 103, 105, 120 n. 1, 121 n. 4
natures 85, 86, 87, 89, 93, 96 n. 14
Nedham, Marchamont 4, 64, 65, 68–76, 77 n. 11, n. 12
Neile, Sir Paul 114
Neile, William 114–120, 126 n. 42, n. 44, 139, 150 n. 16
Newman, L. 96 n. 17, 98 n. 39
Newman, W. R. 77 n. 10
Newton, Sir Isaac 1, 2, 4, 5, 6, 28 n. 10, 80–98 *passim*, 106, 130–51 *passim*, 153, 160, 172, 182–90 *passim*, 215, 225–9, 241; 'Account of the *Commercium Epistolicum*' 98 n. 35; active principles 90, 92, 97 n. 29; alchemical writings 82, 83; and atomism 144; conservation laws 137, 146–8; 'De Gravitatione' 28 n. 10, 81, 82, 86, 88, 95 n. 6, 96 n. 20, 98 n. 35, 132; *De Motu* 80; 'General Scholium' 81, 82, 85, 89, 96 n. 16; on gravity 4, 80–95, 225–9; on hypotheses

85, 94, 227–8; on individuation 147–8; laws of motion 1, 5, 85, 87, 90, 136, 137, 146, 147, 150, n. 12, n. 13, n. 14; letters to Bentley 81, 83, 89–92, 96 n. 22; on mass 83–90, 147, 188 n. 4, n. 5; *Optics* 81, 84, 88, 90, 92, 96 n. 13, 226; pendulum experiments 6, 182–7; on planetary motion 148, 151 n. 22; *Principia* 4, 80, 81, 82–95, 95 n. 6, 96 n. 11, n. 16, n. 20, 97 n. 24, n. 26, n. 27, n. 32, 132, 137, 143, 146, 148, 150, n. 13, n. 20, 151 n. 21, 160; rules of reasoning 83, 87, 88, 93–4, 98 n. 38, 145; theology 81, 83; third law of motion 83, 85, 137, 146–8, 150 n. 14, 151 n. 22, 153, 185, 187; *Treatise of the System of the World* 4, 80–95; *Trinity Notebook* 143
Newton's Challenge 84
Nicole, Pierre 55, 59 n. 10
Nomore, C. 142, 149 n. 2, n. 3

O

occult qualities 63, 108, 228
Oldenburg, Henry 110–20, 124 n. 30, n. 31, n. 33, n. 34, 125 n. 34, n. 35, n. 36, n. 37, n. 39, 126 n. 46, n. 47, 127 n. 48, 161, 169, 170
Olson, R. 187 n. 1

P

Palmerino, C. R. 98 n. 39
Paracelsus 61, 67
particle physics 6, 210
Pascal, Blaise 51, 55, 57, 59 n. 10
pendulum experiments 124 n. 30, 179; of Newton 6, 182–7, 189 n. 20; perfectly elastic 182; perfectly inelastic 179, 186–7
perceptions 87; Hume on 238, 239, 240, 244, 251–4, 257 n. 19; monadic 201, 203–8, 211, 212 n. 19, 230
Pessin, A. 122 n. 17
Phenomenalism 201
physics 5, 6, 11, 105, 112–15, 122 n. 13, 130, 137, 139, 140, 210, 232 n. 8, 244; Aristotelian 227; Cartesian 11, 12, 18, 22, 26, 27 n. 6, 32, 35, 43, 44 n. 3, 50–3, 107–10, 117, 122 n. 17, 123 n. 27, n. 28, 125 n. 37, n. 38, n. 39, 149 n. 2, 240; Leibniz on 220, 225, 227, 228, 230; Newtonian 241; particle physics 6, 210; *see also* mechanics
place, *see* space
Plato 75, 213 n. 24
Platonism 213 n. 25
plenum, *see* void
Pollot, Alphonse 23
Port Royal Logic 55, 58 n. 8, 59 n. 10
Positivism 81
Priest, G. 224, 232 n. 12
primary and secondary quality distinction 1, 6, 66, 235–56 *passim*
primary qualities 6, 66, 92, 235, 237–40, 247–8, 257 n. 16, n. 22; Aristotelian *primae qualitates* 4, 6, 52, 63, 64, 65, 69, 71–2, 76
Principe, L. M. 77 n. 5, n. 10
principles 54, 74, 97 n. 28, 105, 109, 112, 115, 117, 225, 227, 230, 231 n. 4; causal 245, 250, 253; chymical 66, 69–70, 72, 73, 75; corpuscular 66, 68, 71, 72, 76, 77 n. 9; in Descartes 11, 24–5, 31–7, 40, 41, 43, 44, 44 n. 2, n. 3, 45 n. 8, 109, 123 n. 18, 125 n. 38, 131, 145; Galenic 67, 69–73, 75, 76; of mechanics 24, 36, 123 n. 28, 227 of motion 113, 125 n. 38, n. 39, 155, 175; Ockham's razor 56; of plenitude 216; primary qualities as principles 67–8, 76; Primary Qualities Principle 247–9, 257 n. 22, n. 23; Proper Sensibles Principle 236–56, 256 n. 12, n. 16, 257 n. 22, n. 23, n. 24; sufficient reason 224, 225; of unity of composite systems 132, 133–8, 146
proper sensibles 6, 236, 238–44, 247, 250–3, 255, 256 n. 17, 257, n. 23
Ptolemy 32

Q

qualities 1, 2, 4, 51, 52, 109, 110, 120, 123 n. 27, 208; definition of 96 n. 17; relational qualities 80, 83–90, 92; theory of 3, 4, 6,

61–76 *passim*; *see also* primary and secondary quality distinction, primary qualities, secondary qualities, sensible qualities, proper sensibles
queries on motion 107, 112–16, 120, 125 n. 39; Newton's 81, 82, 87–8, 90, 92, 96 n. 13, 232 n. 13

R

Rabener, Gebhard Johann 198
Radelet, P. 150, n. 12
Rationalism 82
Read, R. 235
Realism 82, 95 n. 6, 209, 211, 213 n. 25, 235, 236, 256 n. 3, 257 n. 19
Régis, Pierre-Sylvain 44
Regius, Henricus 44
Reid, Thomas 241–2
Remond, Nicolas 212 n. 12, 213 n. 20
Reneri, Henricus 23
Rescher, N. 231 n. 4, 232 n. 15
resistance: of the air 182–4, 187; to motion 113, 116, 118, 119, 123 n. 18, 125 n. 36, 126 n. 45; *see also* inertia, pendulum experiments
rest 11, 15–26 *passim*, 28 n. 11, n. 13, n. 14, 29 n. 25, 53, 54, 90, 108, 109, 112–19 *passim*, 123 n. 18, 125 n. 37, 147, 150 n. 20, 151 n. 21, 156, 157, 160, 170, 176, 177, 180, 185, 189 n. 12
Richman, K. A. 235
Rochot, Bernard 49
Rohault, Jacques 27 n. 5, 44
Ronan, C. A. 114
Rooke, Lawrence 171
Roux, S. 19, 27 n. 8, 28 n. 19, 142, 150, n. 17
Royal Society 2, 5, 70, 80, 103–27 *passim*; early debates on laws of motion 5, 103–27, 150, n. 12, 154, 155, 169, 170, 171, 172, 175, 182; groupings within 103–6, 110, 122 n. 9
Rozemond, M. 142
rules of reasoning, *see* Newton
rules of motion, *see* laws of motion
Russell, B. 231 n. 3, 232 n. 7, n. 8, 257 n. 27
Rutherford, D. 207, 212 n. 15, 213 n. 22, n. 23

S

Sablière, dame de la (Marguerite Hessein) 55, 56
Schliesser, E. 4, 84, 95 n. 2, n. 3, n. 4, 97 n. 25, 98 n. 34, 257 n. 30
Schmaltz, T. 122 n. 17
secondary qualities 6, 66, 235–55 *passim*; Aristotelian 63
Sedley, D. 256 n. 12
Sennert, Daniel 62, 71
sensible qualities 6, 66, 74, 236, 237–55, 256 n. 8, n. 12, n. 17, 257 n. 16, n. 22
Shackelford, J. 77 n. 13
Shapiro, A. 84
Sleigh, R. C. 212 n. 7
Slowick, E. 122 n. 17
Sluse, René F. 125 n. 34
Smeenk, C. 86, 96 n. 20, 98 n. 39
Smith, Adam 98 n. 33
Smith, G. 81, 96 n. 11, n. 16
Smith, J. E. H. 213 n. 25
Snobelen, S. 82
Socratic Problem 81, 95 n. 3, 97 n. 24
solidity 50–4, 58 n. 3, n. 6, 68, 240; *see also* hardness
Sophie of Hanover, Electress 203
Sophie Charlotte, Queen of Prussia 229
Sorbière, Samuel 104, 105, 121 n. 8, n. 9, 122 n. 9, n. 10, 124 n. 32
souls 18, 38, 141, 142, 195, 196–7, 199, 202–9 *passim*, 210, 212 n. 11, n. 12, 225–6, 219, 229, 230, 231
space 3, 13, 14, 22, 55–7, 59 n. 12, 87, 88, 95 n. 6, 98 n. 35, 109, 110, 111, 124 n. 29, 126 n. 41, 142, 143, 215, 229, 248; absolute 225, 226; Aristotelian 56; *see also* void
Spinoza, Baruch 1, 82, 98 n. 35, 149 n. 11
Sprackling, Robert 4, 69, 72–4, 76
Sprat, Thomas 103–5, 110, 111, 121 n. 2, n. 3, n. 5, n. 7, n. 8, 122 n. 9, n. 10, n. 12
Starkey, George 66, 77 n. 10
states 53, 54, 97 n. 30, 108, 109, 123, 131, 133–4, 137, 147, 148, 149 n. 6, 150, n. 20, 151, n. 21, 160, 185
Stein, H. 80, 83, 85, 92–4, 96 n. 10, n. 13, n. 15, 97 n. 31, 98 n. 36, n. 37, n. 39, 188 n. 4, n. 5
Strawson, G. 256 n. 5

Stubbe, Henry 69
Suárez, Francisco 32, 42
substance 1, 6, 11, 13, 14, 18, 19, 26, 28 n. 17, n. 18, n. 19, n. 22, 55, 95 n. 6, 107, 125 n. 37, 141, 142, 196, 213 n. 21; Hume on 235–40, 247, 248, 255; Leibniz on 196–211, 212 n. 11, n. 12, n. 19, 213 n. 21, 225, 230
substantial form 108, 123 n. 27, 196, 197, 207, 208, 227
superaddition 87, 89, 92, 96 n. 21, 200, 207–8
suppositions 34, 36, 44 n. 1; *see also* hypotheses
Syfret, R. H. 122 n. 9

T

Thomson, George 65, 70, 77 n. 6, n. 13
thoracic duct 62
Torricelli, Evangelista 175
transduction 24, 29 n. 29
Turnbull, H. W. 97 n. 32
Twysden, John 4, 69, 74–6, 77 n. 16, n. 17
Tycho, *see* Brahe

U

universe: worlds within worlds model 6, 221–4

V

Valesius, Franciscus 74
Van Dyck, M. 98 n. 39
Van Lunteren, F. 98 n. 33
Verelst, K. 98 n. 39
Virgil 242, 243
Vogel, Martin 125 n. 34
void, vacuum 13, 14, 27 n. 7, 29 n. 30, 36, 51–7, 91, 97 n. 28, 120, 124 n. 29, 182–5; interstitial spaces 52, 53, 54; Leibniz on 215–31
Voltaire 121 n. 9

W

Wadham College, 114
Wagner, Gabriel 212 n. 10, n. 11
Wallis, John 5, 6, 106, 107, 110–15, 119, 124 n. 33, 125 n. 34, n. 36, n. 37, n. 39, 126 n. 42, 153, 154, 155, 189 n. 18; collision laws 175–81, 182, 187
Walmsley, J. C. 77 n. 5, n. 13
Ward, Seth 114
weapon salve 66
Wear, A. 76 n. 2
weight 13, 14, 93, 157, 160, 175–7, 180, 188 n. 4, 256 n. 12; medieval science of weights 179
Westfall, R. S. 150, n. 13, n. 14
Wilkins, John 114
Willis, Thomas 69, 70, 77 n. 9, n. 13
Wilson, C. 5, 139, 154, 156, 161, 171, 175, 187 n. 1, 189 n. 14
Winkler, K. 257 n. 30
Winterbourne, A. T. 232 n. 15
Wolff, Christian 213 n. 24
Wood, P. 121 n. 4
Woolhouse, R. S. 122 n. 16
Wren, Sir Christopher 5, 6, 105–7, 110–16, 121 n. 8, 124 n. 30, n. 31, n. 33, 125 n. 34, n. 35, 150, n. 12, 154, 155, 169–73; collision laws 170–4, 182, 186, 187, 189 n. 14
Wright, J. 247, 251, 256 n. 5, 257 n. 19, n. 25

BLOODLINES
A DYING TRUTH EXPOSED, BOOK ONE

HER JOURNEY IS ONLY THE BEGINNING OF A LEGACY

MARCUS ABSTON

Bloodlines (A Dying Truth Exposed, Book One)
Copyright © 2020 by Marcus Abston, Chas Novels

All rights reserved. No part of this publication may be reproduced, distributed, or transmitted in any form or by any means, including photocopying, recording, or other electronic or mechanical methods, without the prior written permission of the publisher, except in the case of brief quotations embodied in critical reviews and certain other noncommercial uses permitted by copyright law.

Editing by The Pro Book Editor
Interior and Cover Design by IAPS.rocks

ISBN: 978-1-0879-2458-8

    1. Main category—Fiction
    2. Other category—Historical Fiction

First Edition